Communications
in Computer and Information Science   1223

Commenced Publication in 2007
Founding and Former Series Editors:
Simone Diniz Junqueira Barbosa, Phoebe Chen, Alfredo Cuzzocrea,
Xiaoyong Du, Orhun Kara, Ting Liu, Krishna M. Sivalingam,
Dominik Ślęzak, Takashi Washio, Xiaokang Yang, and Junsong Yuan

More information about this series at http://www.springer.com/series/7899

Alexander Elizarov · Boris Novikov ·
Sergey Stupnikov (Eds.)

Data Analytics
and Management
in Data Intensive Domains

21st International Conference, DAMDID/RCDL 2019
Kazan, Russia, October 15–18, 2019
Revised Selected Papers

 Springer

Editors
Alexander Elizarov
Kazan Federal University
Kazan, Russia

Boris Novikov ⓘ
National Research University
Higher School of Economics
St. Petersburg, Russia

Sergey Stupnikov ⓘ
Federal Research Center
"Computer Science and Control"
Moscow, Russia

ISSN 1865-0929 ISSN 1865-0937 (electronic)
Communications in Computer and Information Science
ISBN 978-3-030-51912-4 ISBN 978-3-030-51913-1 (eBook)
https://doi.org/10.1007/978-3-030-51913-1

This Springer imprint is published by the registered company Springer Nature Switzerland AG
The registered company address is: Gewerbestrasse 11, 6330 Cham, Switzerland

Preface

This CCIS volume published by Springer contains the post-proceedings of the XXI International Conference on Data Analytics and Management in Data Intensive Domains (DAMDID/RCDL 2019) that took place during October 15–18 at the Kazan Federal University, Russia.

DAMDID is held as a multidisciplinary forum of researchers and practitioners from various domains of science and research, promoting cooperation and exchange of ideas in the area of data analysis and management in domains driven by data-intensive research. Approaches to data analysis and management being developed in specific data-intensive domains (DID) of X-informatics (such as X = astro, bio, chemo, geo, med, neuro, physics, chemistry, material science, etc.), social sciences, as well as in various branches of informatics, industry, new technologies, finance, and business are expected to contribute to the conference content.

Traditionally DAMDID/RCDL proceedings are published locally before the conference as a collection of full texts of all contributions accepted by the Program Committee: regular and short papers, abstracts of posters, and demos. Soon after the conference, the texts of regular papers presented at the conference are submitted for online publishing in a volume of the European repository of the CEUR workshop proceedings, as well as for indexing the volume content in DBLP and Scopus. Since 2016, a DAMDID/RCDL volume of post-conference proceedings with up to one third of the submitted papers which have been previously published before in CEUR workshop proceedings are being published with Springer in their *Communications in Computer and Information Science* (CCIS) series. Each paper selected for the CCIS post-conference volume should be modified as follows: the title of each paper should be a new one; the paper should be significantly extended (with at least 30% new content); and the paper should refer to its original version in the CEUR workshop proceedings.

The program of DAMDID/RCDL 2019 was oriented towards data science and data-intensive analytics as well as on data management topics. The program of this year included keynotes and invited talks covering a broad range of conference topics.

The keynote by Bernhard Thalheim (full professor at Christian-Albrechts-University at Kiel) is devoted to models as one of the universal instruments of humans. Four specific utilization scenarios for models are considered as well as methodologies for the development of proper and well-applicable models. Anton Osokin (leading research fellow and deputy head of the Centre of Deep Learning and Bayesian Methods at National Research University Higher School of Economics) gave a talk on ways for combining neural networks and powerful algorithms already developed in various domains. Experiences in creating B.Sc. and M.Sc. study programs in Data Science were considered in the keynote by Ivan Luković (full professor at the Faculty of Technical Sciences, University of Novi Sad). The invited talk by Mikhail Zymbler (head of the Department for Data Mining and Virtualization at South Ural State

University) considered an approach for the integration of data mining methods with open-source relational DBMS enhanced by small-scale modifications of the original codes to encapsulate parallelism.

The workshop on Data and Computation for Materials Science and Innovation (DACOMSIN) hosted by the National University of Science and Technology MISiS in Moscow constituted the first day of the conference on October 15. The workshop was aimed to address the communication gap across communities in the domains of materials data infrastructures, materials data analysis, and materials in silico experiment. The workshop brought together professionals from across research and innovation to share their experience and perspectives of using information technology and computer science for materials data management, analysis, and simulation.

The conference Program Committee reviewed 34 submissions for the conference, 18 submissions for DACOMSIN workshop, and 5 submissions for the PhD workshop. For the conference, 17 submissions were accepted as full papers, 10 as short papers, whereas 7 submissions were rejected. For the DACOMSIN Workshop, 3 submissions were accepted as full papers, 3 submissions were accepted as short papers, 5 submissions were accepted as demos, and 7 submissions were accepted as posters. For the PhD workshop, 4 papers were accepted and 1 was rejected.

According to the conference and workshops program, these 42 oral presentations were structured into 11 sessions, including: Advanced Data Analysis Methods; Digital Libraries and Data Infrastructures; Data Integration, Ontologies and Applications; Data and Knowledge Management; Data Analysis in Astronomy; Information Extraction from Text; Data Formats; Metadata and Ontologies in Support of Materials Research; Materials Databases, Materials Data Infrastructures and Data Services; IT Applications; and IT Platforms for Materials Design and Simulation.

Though most of the presentations were dedicated to the results of researches conducted in the research organizations located on the territory of Russia, including: Chernogolovka, Dubna, Ekaterinburg, Kazan, Moscow, Novosibirsk, Obninsk, Tomsk, Samara, and St. Petersburg, the conference featured international works of seven talks prepared by the foreign researchers from countries such as Armenia (Yerevan), China (Shenzhen, Beijing), France (Clermont-Ferrand), the UK (Harwell), Japan (Tokyo), and Switzerland (Lausanne).

For the post-conference proceedings 15 papers were selected by the Program Committee (11 peer reviewed and 4 invited papers) and after careful editing they formed the content of the post-conference volume structured into 7 sections including Advanced Data Analysis Methods (2 papers), Data Infrastructures and Integrated Information Systems (2 papers), Models, Ontologies and Applications (2 papers), Data Analysis in Astronomy (3 papers), Information Extraction from Text (4 papers), Distributed Computing (1 paper), and Data Science for Education (1 paper).

The chairs of Program Committee express their gratitude to the Program Committee members for carrying out the reviewing of the submissions and selection of the papers for presentation, to the authors of the submissions, as well as to the host organizers from Kazan Federal University. The Program Committee appreciates the possibility of using the Conference Management Toolkit (CMT) sponsored by Microsoft Research, which provided great support during various phases of the paper submission and reviewing process.

April 2020   Alexander Elizarov
                                                                      Boris Novikov
                                                                   Sergey Stupnikov

Organization

General Chair

Alexander Elizarov   Kazan Federal University, Russia

Program Committee Co-chairs

Boris Novikov   National Research University Higher School
of Economics, Russia

Sergey Stupnikov   Federal Research Center Computer Science
and Control, RAS, Russia

DACOMSIN Workshop Co-chairs

Nadezhda Kiselyova   IMET, RAS, Russia

Vasily Bunakov   Science and Technology Facilities Council, UK

PhD Workshop Chair

Mikhail Zymbler   South Ural State University, Russia

Organizing Committee Chair

Ayrat Khasyanov   Kazan Federal University, Russia

Organizing Committee Deputy Chair

Denis Zuev   Kazan Federal University, Russia

Organizing Committee

Iurii Dedenev   Kazan Federal University, Russia

Elena Tutubanilna   Kazan Federal University, Russia

Evgeny Lipachev   Kazan Federal University, Russia

Nikolay Skvortsov   Federal Research Center Computer Science
and Control, RAS, Russia

Victor Zakharov   Federal Research Center Computer Science
and Control, RAS, Russia

Natalya Zaitseva   Kazan Federal University, Russia

Supporters

Kazan Federal University, Russia
Federal Research Center Computer Science and Control of the Russian Academy
    of Sciences (FRC CSC RAS), Russia
Moscow ACM SIGMOD Chapter, Russia

Coordinating Committee

Igor Sokolov (Co-chair)	Federal Research Center Computer Science and Control, RAS, Russia
Nikolay Kolchanov (Co-chair)	Institute of Cytology and Genetics, SB RAS, Russia
Sergey Stupnikov (Deputy Chair)	Federal Research Center Computer Science and Control, RAS, Russia
Arkady Avramenko	Pushchino Radio Astronomy Observatory, RAS, Russia
Pavel Braslavsky	Ural Federal University, SKB Kontur, Russia
Vasily Bunakov	Science and Technology Facilities Council, UK
Alexander Elizarov	Kazan Federal University, Russia
Alexander Fazliev	Institute of Atmospheric Optics, RAS, Siberian Branch, Russia
Alexei Klimentov	Brookhaven National Laboratory, USA
Mikhail Kogalovsky	Market Economy Institute, RAS, Russia
Vladimir Korenkov	JINR, Russia
Mikhail Kuzminski	Institute of Organic Chemistry, RAS, Russia
Sergey Kuznetsov	Institute for System Programming, RAS, Russia
Vladimir Litvine	Evogh Inc., USA
Archil Maysuradze	Moscow State University, Russia
Oleg Malkov	Institute of Astronomy, RAS, Russia
Alexander Marchuk	Institute of Informatics Systems, RAS, Siberian Branch, Russia
Igor Nekrestjanov	Verizon Corporation, USA
Boris Novikov	Saint Petersburg State University, Russia
Nikolay Podkolodny	ICaG, SB RAS, Russia
Aleksey Pozanenko	Space Research Institute, RAS, Russia
Vladimir Serebryakov	Computing Center of RAS, Russia
Yury Smetanin	Russian Foundation for Basic Research, Russia
Vladimir Smirnov	Yaroslavl State University, Russia
Konstantin Vorontsov	Moscow State University, Russia
Viacheslav Wolfengagen	National Research Nuclear University MEPhI, Russia
Victor Zakharov	Federal Research Center Computer Science and Control, RAS, Russia

Program Committee

Alexander Afanasyev	Institute for Information Transmission Problems, RAS, Russia
Arkady Avramenko	Pushchino Observatory, Russia
Ladjel Bellatreche	National Engineering School for Mechanics and Aerotechnics, France
Pavel Braslavski	Ural Federal University, Russia
Vasily Bunakov	Science and Technology Facilities Council, UK
Evgeny Burnaev	Skolkovo Institute of Science and Technology, Russia
Yuri Demchenko	University of Amsterdam, The Netherlands
Jerome Darmont	ERIC, Université Lumière Lyon 2, France
Boris Dobrov	Research Computing Center, MSU, Russia
Alexander Elizarov	Kazan Federal University, Russia
Alexander Fazliev	Institute of Atmospheric Optics, SB RAS, Russia
Yuriy Gapanyuk	Bauman Moscow State Technical University, Russia
Vladimir Golenkov	Belarusian State University of Informatics and Radioelectronics, Belarus
Vladimir Golovko	Brest State Technical University, Belarus
Olga Gorchinskaya	FORS Group, Russia
Evgeny Gordov	Institute of Monitoring of Climatic and Ecological Systems, SB RAS, Russia
Valeriya Gribova	Institute of Automation and Control Processes, FEB RAS, Far Eastern Federal University, Russia
Maxim Gubin	Google Inc., USA
Natalia Guliakina	Belarusian State University of Informatics and Radioelectronics, Belarus
Sergio Ilarri	University of Zaragoza, Spain
Mirjana Ivanovic	University of Novi Sad, Serbia
Nadezhda Kiselyova	IMET, RAS, Russia
Mikhail Kogalovsky	Market Economy Institute, RAS, Russia
Sergey Kuznetsov	Institute for System Programming, RAS, Russia
Dmitry Lande	Institute for Information Recording, NASU, Ukraine
Evgeny Lipachev	Kazan Federal University, Russia
Giuseppe Longo	University of Naples Federico II, Italy
Natalia Loukachevitch	Lomonosov Moscow State University, Russia
Ivan Lukovic	University of Novi Sad, Serbia
Oleg Malkov	Institute of Astronomy, RAS, Russia
Yannis Manolopoulos	Open University of Cyprus, Cyprus
Archil Maysuradze	Lomonosov Moscow State University, Russia
Manuel Mazzara	Innopolis University, Russia
Alexey Mitsyuk	National Research University Higher School of Economics, Russia
Xenia Naidenova	S. M. Kirov Military Medical Academy, Russia
Dmitry Namiot	Lomonosov Moscow State University, Russia
Igor Nekrestyanov	Verizon Corporation, USA

Panos Pardalos	University of Florida, USA
Nikolay Podkolodny	Institute of Cytology and Genetics, SB RAS, Russia
Jaroslav Pokorny	Charles University in Prague, Czech Republic
Natalia Ponomareva	Scientific Center of Neurology, RAMS, Russia
Alexey Pozanenko	Space Research Institute, RAS, Russia
Andreas Rauber	Vienna Technical University, Austria
Timos Sellis	Swinburne University of Technology, Australia
Vladimir Serebryakov	Computing Centre, RAS, Russia
Nikolay Skvortsov	Federal Research Center Computer Science and Control, RAS, Russia
Manfred Sneps-Sneppe	AbavaNet, Latvia
Leonid Sokolinskiy	South Ural State University, Russia
Valery Sokolov	Yaroslavl State University, Russia
Alexander Sychev	Voronezh State University, Russia
Bernhard Thalheim	University of Kiel, Germany
Alexey Ushakov	University of California, Santa Barbara, USA
Dmitry Ustalov	University of Mannheim, Germany
Pavel Velikhov	Finstar Financial Group, Russia
Alexey Vovchenko	Federal Research Center Computer Science and Control, RAS, Russia
Peter Wittenburg	Max Planck Computing and Data Facility, Germany
Anna Yarygina	Saint Petersburg State University, Russia
Vladimir Zadorozhny	University of Pittsburgh, USA
Yury Zagorulko	Institute of Informatics Systems, SB RAS, Russia
Victor Zakharov	Federal Research Center Computer Science and Control, RAS, Russia, Russia
Sergey Znamensky	Institute of Program Systems, RAS, Russia

DACOMSIN Workshop Program Committee

Toshihiro Ashino	Toyo University, Japan
Rossella Aversa	Karlsruhe Institute of Technology, Germany
Vasily Bunakov	Science and Technology Facilities Council, UK
Stefano Cozzini	CNR-IOM Democritos, Italy
Martin Horsch	STFC Daresbury Laboratory, UK
Nadezhda Kiselyova	IMET, RAS, Russia
Igor Morozov	Joint Institute for High Temperatures, RAS, Russia
Björn Schembera	High Performance Computing Center Stuttgart (HLRS), Germany
Leopold Talirz	EPFL, Switzerland
Irina Uspenskaya	Lomonosov Moscow State University, Russia
Yibin Xu	National Institute for Materials Science, Japan

PhD Workshop Program Committee

Alexander Fazliev	Institute of Atmospheric Optics, SB RAS, Russia
Yuriy Gapanyuk	Bauman Moscow State Technical University, Russia
Sachin Kumar	South Ural State University, Russia
Sergey Kuznetsov	Institute for System Programming, RAS, Russia
Natalia Loukachevitch	Lomonosov Moscow State University, Russia
Ivan Lukovic	University of Novi Sad, Serbia
Alexey Mitsyuk	National Research University Higher School of Economics, Russia
Dmitry Nikitenko	Lomonosov Moscow State University, Russia
Roman Samarev	Bauman Moscow State Technical University, Russia
Sergey Sobolev	Lomonosov Moscow State University, Russia
Sergey Stupnikov	FRC CSC RAS, Russia
Pavel Velikhov	Huawei, Russia

Contents

Information Extraction from Text

Distributed Computing

Data Science for Education

Advanced Data Analysis Methods

Three Simple Approaches to Combining Neural Networks with Algorithms

Anton Osokin$^{(\boxtimes)}$ (iD)

National Research University Higher School of Economics, Moscow, Russia
`aosokin@hse.ru`

Abstract. Recently, deep neural networks have showed amazing results in many fields. To build such networks, we usually use layers from a relatively small dictionary of available modules (fully-connected, convolutional, recurrent, etc.). Being restricted with this set of modules complicates transferring technology to new tasks. On the other hand, many important applications already have a long history and successful algorithmic solutions. Is it possible to use existing methods to construct better networks? In this paper, we cover three approaches to combining neural networks with algorithms and discuss their pros and cons. Specifically, we will discuss three approaches: structured pooling, unrolling of algorithm iterations into network layers and explicit differentiation of the output w.r.t. the input.

Keywords: Deep learning · Algorithms · Backpropagation

1 Introduction

Recently, neural networks have achieved remarkable success in many fields, and many practical systems for fundamental tasks are built with neural networks. For example, in computer vision, it is image classification [16,20], object detection [14,32] and image segmentation [26,33]; in natural language processing, it is language modeling [12,13] and automatic translation [4,39]; in audio processing, both speech recognition [17] and synthesis [37]. Many approaches have become an industrial standard, and companies around the world are building products based on this technology.

Successful algorithms for various tasks are very different from each other and required years of research to arrive at the current level of performance. Constructing a good algorithm for a new task is often a non-trivial challenge. It also turns out that networks can not just learn from data without exploiting some domain knowledge. This knowledge is usually encoded at least in the architecture of the network. For example, convolutional neural networks [15] exploit intuition that translation of the object does not change the object, i.e., a cat does not stop being a cat if moved left.

At the same time, in many domains, we already have powerful algorithms that do a decent job. It is a very natural idea to exploit those to construct

© Springer Nature Switzerland AG 2020
A. Elizarov et al. (Eds.): DAMDID/RCDL 2019, CCIS 1223, pp. 3–12, 2020.
https://doi.org/10.1007/978-3-030-51913-1_1

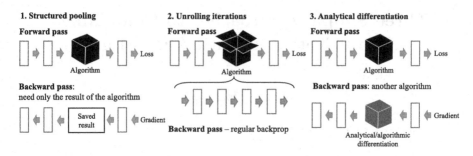

Fig. 1. Thee approaches to combine an algorithm and a neural network.

better networks. We can thinks about this from the two sides: we use algorithms to construct new layers for networks that encode domain knowledge or make existing algorithms trainable and adjust them to data. In any case, the attempt is to take best of both worlds. This direction has been around since 90s [7,22,24], but for long time was not getting significant attention (together with neural networks).

In this paper, we will review three ways to combine algorithms and networks (see Fig. 1):

1. structured pooling: an algorithm is used to select active features (similarly to max pooling), when computing the gradient one needs only the result of the algorithm;
2. unrolling iterations into layers: an algorithm becomes a part of the network;
3. explicit derivative w.r.t. the algorithm input, i.e., building a layer with an explicit implementation of the derivative computation.

This paper is an extended version of the abstract [29].

2   Example Task: Handwriting Recognition

To illustrate the first two approaches, we will use an example of a simplified task of handwriting recognition: recognize a word given a sequence of images where each image shows exactly one letter. See input/output example below.

We will work with the OCR dataset of Taskar et al. [35],[1] which can be viewed as the MNIST of structured prediction. Importantly, to achieve good quality on this dataset, one has to use take the whole input sequence into account: a classifier, e.g., ConvNet, trained on individual letters delivers around 8% of errors, because some letters are not recognizable individually; the models that consider sequences deliver 1–3% errors (see examples in [23,31,34]).

---

[1] http://ai.stanford.edu/~btaskar/ocr/.

In this paper, we use the standard convolutional network (ConvNet) for MNIST as a letter classifier [22] (we substitute the last layer with 10 outputs for digits with the similar layer with 26 outputs for letters). This network consists of linear layers with parameters (convolutional and fully-connected) intervened by nonlinearities and pooling operations. The model is trained by stochastic optimization of the cross-entropy loss, where the gradient is computed by back-propagation. See, e.g., the PyTorch tutorial[2] for a complete example.

Notation. More formally, we will work in the standard supervised setting, where at the learning stage, we are given a dataset $\{x_i, y_i\}_{i=1}^{N}$ where each item is represented by an input sequence of individual images $x_i := \{x_{i,t}\}_{t=1}^{T}$ and a target sequence of class labels $y_i := \{y_{i,t}\}_{t=1}^{T}$ from a finite alphabet $y_{i,t} \in 1, \ldots, K$ (for English letters, we have $K = 26$). For simplicity and without loss of generality, we will assume that all the sequences are of the same length T.

The most direct approach to recognize sequences of letter images is to classify each image independently. For each image t of sequence i, we can apply a ConvNet classifier f_W : letter image $\to \mathbb{R}^K$ to get a K-dimensional vector of scores showing compatibility with each of the K classes. The symbol W denotes all the learnable parameters of the network, which in this case, are the kernels and biases of convolutional layers and weight matrices and biases of fully-connected layers. We will use the notation $f_W(y \mid x_{i,t})$ to refer to the score of class y for image $x_{i,t}$. With this model, a sequence of images $x = \{x_t\}_{t=1}^{T}$ can be classified by finding the maximum joint score defined as the sum of scores over the sequence:

$$\left\{ \underset{y}{\mathrm{argmax}}\, f_W(y \mid x_t) \right\}_{t=1}^{T} = \underset{y_1,\ldots,y_T}{\mathrm{argmax}} \sum_{t=1}^{T} f_W(y_t \mid x_t). \tag{1}$$

For each sequence of score vectors $F := \{f_t\}_{t=1}^{T}$, $f_t \in \mathbb{R}^K$, $f_{t,y} := f_W(y \mid x_{i,t})$ with corresponding labels $y = \{y_t\}_{t=1}^{T}$, we can define a sequence loss function as the sum of the regular cross-entropy losses (often referred to as the log loss) over the T sequence positions:

$$\ell(F, y) := \sum_{t=1}^{T} \left(-f_{t,y_t} + \log \sum_{k=1}^{K} \exp f_{t,k} \right). \tag{2}$$

The training process consists in stochastic minimization of the sequence loss functions average over the dataset. Each training iteration consists of several steps:

1. randomly choose an item $x_i = \{x_{i,t}\}_{t=1}^{T}$ with annotation $y_i = \{y_{i,t}\}_{t=1}^{T}$;
2. perform the network forward pass to get $F_i := \{f_W(x_{i,t})\}_{t=1}^{T}$;
3. compute the loss $\ell(F_i, y_i)$ based on the network outputs F_i and the annotation y_i;
4. differentiate the loss w.r.t. the network output, i.e., compute $\frac{d\ell}{dF}$;
5. obtain the gradient w.r.t. the parameters $\frac{d\ell}{dW}$ by back-propagating $\frac{d\ell}{dF}$ into the network;

---

[2] https://github.com/pytorch/examples/blob/master/mnist/main.py.

6. perform one step of a stochastic optimization algorithms, e.g., SGD or Adam.

Note that the modern deep learning libraries such as TensorFlow [1] and PyTorch [30] perform automatic differentiation, which means that steps 4 and 5 are automatic, and provide a lot of tools to help implementing other steps.

Also note that the network usually contains functions that are not differentiable everywhere (e.g., the max function or rectified-linear unit, ReLU), but as typical in deep learning we will ignore the non-differentiability and just use some meaningful direction if the training process arrives at one of those points (for example, we will break the ties of the max function in some deterministic way, i.e., use $\frac{df}{dx}$ as an approximation of $\frac{d}{dx} \max(f(x), g(x))$ at the points where $f(x) = g(x)$).

3 Structured Pooling

Within the structured pooling approach to combining networks and algorithms, we run the forward pass of the network, and then run our algorithm as a black-box function. For the backward pass, we only need the result of the algorithm and not the algorithm itself. In such settings, the algorithm implementation can be independent from the network implementation. One can call algorithms implemented in different libraries and possibly even run on different hardware.

We will give an example of structured pooling in the context of our OCR example (see Sect. 2). To model relations between letters, we will use a conditional random field (CRF) model [21]. This model is defined by a score function which assigns a score (a real number) to each labeling of all the variables (all letter sequences of the correct length). To be able to deal with the score function efficiently, we need it to be of some special form. The simplest way to go is to use a function that consists of unary and pairwise potentials:

$$F(\boldsymbol{y} \mid \boldsymbol{\theta}) := \sum_{t=1}^{T} \theta_t^U(y_t) + \sum_{t=1}^{T-1} \theta_t^P(y_t, y_{t+1}). \tag{3}$$

The unary potentials $\theta_t^U(y_t)$ represent information about individual letters and we will simply set them to the output of the neural network $f_W(y_t \mid x_t)$. The pairwise potentials $\theta^P(y_t, y_{t+1})$ represent connections between neighboring symbols, and we will parameterize them with a single matrix $\Theta^P \in \mathbb{R}^{K \times K}$. The score function (3) represents a chain-like graphical model (Hidden Markov Model) and a lot of operations can be done exactly in polynomial time [5, Section 13.2].

Now we move to training the system defined by the score function (3). The set of all the parameters include the parameters of the ConvNet that defines the unary potentials and in the matrix Θ^P defining the pairwise potentials. Recall the standard supervised setting with the training set $\{\boldsymbol{x}_i, \boldsymbol{y}_i\}_{i=1}^{N}$ where each item is represented by an input sequence of individual images $\boldsymbol{x}_i := \{x_{i,t}\}_{t=1}^{T}$ and the target sequence of class labels $\boldsymbol{y}_i := \{y_{i,t}\}_{t=1}^{T}$, $y_{i,t} \in 1, \dots, K$. For the training objective, a common choice is the Structured SVM approach [35,36], SSVM, which is also called structured max-margin:

$$\ell_{\text{SSVM}}(F, \boldsymbol{y}) = \max_{\hat{\boldsymbol{y}}} \left(F(\hat{\boldsymbol{y}} \mid \boldsymbol{\theta}) + \Delta(\hat{\boldsymbol{y}}, \boldsymbol{y}) \right) - F(\boldsymbol{y} \mid \boldsymbol{\theta}). \tag{4}$$

The function $\Delta(\hat{y}, y)$ defines the margin between the annotation y and an argmax configuration $\hat{y}$, which will both have a high score value and be far from the annotation.

Note that the maximization operation in (4) requires searching over K^T configurations, which is usually intractable for exhaustive search. However, the structure of the score function (3) allows to efficiently find the maximum by the dynamic programming principle, the algorithm is known as the Viterbi algorithm in this case. The algorithm is very simple and consists in iteratively computing the Bellman function (also called the DP table). After the computation, we can easily restore the argmax configuration using the positions of a maximal elements appearing in Bellman iterations (also known as parent pointers). See, e.g., [5, Section 13.2.5] for the detailed description.

Each iteration of the training process will consist of several steps:

1. randomly choose an item $x_i = \{x_{i,t}\}_{t=1}^T$ with annotation $y_i = \{y_{i,t}\}_{t=1}^T$;
2. perform the network forward pass to get $\theta_t^U := f_W(x_{i,t})$;
3. solve (or approximate) the maximization problem $\max_{\hat{y}}\left(F(\hat{y} \mid \theta) + \Delta(\hat{y}, y)\right)$, save an argmax configuration $\hat{y}$;
4. compute the loss ℓ_{SSVM} based on the score function F, the annotation y_i and the obtained $\hat{y}$;
5. differentiate the loss (assuming it does not contain the max operator and equals $F(\hat{y} \mid \theta) + \Delta(\hat{y}, y) - F(y \mid \theta)$) w.r.t. the network outputs f_W;
6. obtain the gradient w.r.t. the parameters $\frac{d\ell}{dW}$ by back-propagating $\frac{d\ell}{dF}$ into the network;
7. perform one step of a stochastic optimization algorithms, e.g., SGD or Adam.

Applying the SSVM loss (4) results in an instance of structured pooling because step 5 requires only the argmax vector $\hat{y}$ computed at step 3, and, for the rest of the method, it is not important how exactly $\hat{y}$ was computed. Most importantly, we never need to backpropagate gradients through the internal operations of the dynamic programming algorithm, but only use its results.

Speaking about applications, methods built on the structured pooling approach were successfully applied in multiple computer vision tasks, for example, free text recognition [19], joint detection of multiple objects [38] and image tagging [10].

4 Unrolling Iterations

By unrolling iterations, we refer to an approach where we open the black box of the algorithm, take its iterations and literally plug those into a network as layers. The gradient of the whole system can be computed using automatic differentiation packages.

As an example of this approach we will look at the maximum likelihood method of training a handwriting recognition system. Consider the linear-chain score function (3) defined in Sect. 3. We now define the probability distribution over all possible configurations as follows:

$$P(\boldsymbol{y} \mid \boldsymbol{\theta}) = \tfrac{1}{Z(\boldsymbol{\theta})} \exp\big(F(\boldsymbol{y} \mid \boldsymbol{\theta})\big), \quad Z(\boldsymbol{\theta}) = \sum_{\hat{y}} \exp\big(F(\hat{\boldsymbol{y}} \mid \boldsymbol{\theta})\big). \tag{5}$$

Here $Z(\boldsymbol{\theta})$ is the normalization constant (also known as the partition function), and computing it requires summing over K^T summands, which is usually intractable. However, the chain-like structure of the score function (3) allows constructing an efficient algorithm: similarly to the maximization within the SSVM training, we will use a version of dynamic programming. The algorithm is known as the sum-product message passing (a.k.a. the forward-backward or Baum-Welch algorithm). See [5, Section 13.2.2] for the details. Importantly, this algorithm only contains differentiable operations such as the sum, product and exponent, thus automatic differentiation allows to easily compute $\frac{dZ}{d\theta}$.

Taking the negative logarithm of the probability function (5) we arrive at the negative log-likelihood loss function (NLL)

$$\ell_{\text{NLL}}(F, \boldsymbol{y}) := -\log P(\boldsymbol{y} \mid \boldsymbol{\theta}) = -F(\boldsymbol{y} \mid \boldsymbol{\theta}) + \log Z(\boldsymbol{\theta}). \tag{6}$$

The overall training scheme almost exactly matches the scheme described in Sect. 3. However there are two important differences:

– the algorithm has to be implemented in the same system as the network and should contain only the operations supported by the automatic differentiation engine;
– the results of all the intermediate steps of dynamic programming have to be stored in memory.

Note that the network operation are often run on specialized devices, e.g., GPUs, and the algorithm on top is run on the same devices. Storing all the intermediate steps in the device memory can easily become a bottleneck of the whole system.

The approach described in this section was used in many applications including image segmentation [40], image denoising [25], hyper-parameter optimization [27], meta-learning [3] and adversarial generative models [28].

5   Explicit Derivatives w.r.t the Input

In this section, we discuss the last approach of differentiating algorithms. We will illustrate the approach on the Gaussian CRF model proposed by Chandra et al. [8,9] for the task of image segmentation. For the sake of simplicity, we will assume that the task has only two labels of each variable ($K = 2$). The joint score function will be defined as a quadratic function

$$F(\boldsymbol{y} \mid \boldsymbol{\theta}, A) := -\tfrac{1}{2}\boldsymbol{y}^{\top} A \boldsymbol{y} + \boldsymbol{y}^{\top} \boldsymbol{\theta}, \tag{7}$$

where $\boldsymbol{y} \in \mathbb{R}^T$ is a vector of real-valued variables (correspond to image pixels in the context of image segmentation). When the matrix A is positive definite the maximization of the score F is equivalent to solving the system of linear equations $Ay = \theta$ or computing $y = A^{-1}\theta$.

Now consider a layer $y(\theta, A) := \text{argmax}_y F(y \mid \theta, A)$. The task of differentiation of this layer consists in computing $\frac{d\ell}{d\theta}$ and $\frac{d\ell}{dA}$ given the derivative of the loss function $\frac{d\ell}{dy}$ w.r.t. the layer output. We can use the standard matrix calculus to get $\frac{d\ell}{d\theta} = A^{-\top}\frac{d\ell}{dy}$ and $\frac{d\ell}{dA} = -\frac{d\ell}{d\theta}y^\top$, which allow to implement the derivatives explicitly.

Importantly, the process described above does not rely on unrolling the iterations of the solver used at the forward pass, thus can easily use external solvers. The main drawback of this approach is that one needs to derive and implement the backward operations explicitly and cannot rely on the power of automatic differentiation. However, recently there have been many works suggesting (or finding from the literature) methods to differentiate many important algorithms. Ionescu et al. [18] provided the derivatives for several linear algebra algorithms including SVD, Chen et al. [11] used a neural network containing a solver for ordinary differential equations. Finally the line of work by Agrawal et al. [2] used the technique of implicit differentiation to differentiate solvers for constrained convex optimization.

6 Conclusion

In this paper, we reviewed three approaches to combining algorithms and neural networks: structured pooling, unrolling iterations and explicit differentiation. Explicit differentiation is probably the best approach, but it requires to manually design and implement algorithms to compute derivatives, thus has limited applicability. Most often it is used to deal with linear algebra operations [18], but was applied with several other domains such as differential equations [11] and convex optimization [2]. Structured pooling still allows to use out-the-box implementations of the algorithms at the forward pass, but does not need a second algorithm for the backward pass. However, it puts heavy requirements on how the algorithm results are used: the algorithm should be selecting a subset of the features, which can often lead to sparse gradients (not good for training). Unrolling the algorithm iterations is probably the most general and default approach of today, but it makes extensive use of the GPU memory which is often the limiting resource. It is also not clear how to choose the number of iterations for iterative algorithms.

For an experimental demonstration of how algorithms can improve performance of neural networks, we refer to the work of Shevchenko and Osokin [34]. For convenience of the readers, we restate their results in Tables 1 and 2 for the tasks of handwriting recognition (see Sect. 2) and binary segmentation (the Weizmann Horses dataset [6] with the UNet model [33] as the unary network), respectively.

Table 1. Results on the OCR dataset: comparing a classifier of individual digits against two structured approaches.

Method	Accuracy
NN: symbol classifier (2)	91.8 ± 0.2
Structured pooling: SSVM (4)	96.4 ± 0.4
Unrolling iterations: NLL (5)	96.5 ± 0.3

Table 2. Results on the Weizmann Horses dataset: comparing a classifier of individual pixels (with shared features) digits against a structured approach.

Method	Jaccard index
NN: UNet (2)	84.1 ± 0.2
Explicit Derivatives: GCRF (7)	85.6 ± 0.3

Besides illustrating the benefits of embedding some algorithms into neural networks, the work [34] raised several issues with these approaches. Since backpropagation is a leaky abstraction, even if we can compute the gradient it does not mean that the system will train successfully. See the work [34] for the discussion of specific challenges of training and some possible solutions.

Acknowledgements. This work was supported by the Russian Science Foundation project 19-71-00082.

References

1. Abadi, M., Agarwal, A., Barham, P., et al.: TensorFlow: large-scale machine learning on heterogeneous systems (2015). http://tensorflow.org/
2. Agrawal, A., Amos, B., Barratt, S., Boyd, S., Diamond, S., Kolter, Z.: Differentiable convex optimization layers. In: Advances in Neural Information Processing Systems 32 (NeuIPS) (2019)
3. Andrychowicz, M., et al.: Learning to learn by gradient descent by gradient descent. In: Advances in Neural Information Processing Systems 29 (NIPS) (2016)
4. Bahdanau, D., Cho, K., Bengio, Y.: Neural machine translation by jointly learning to align and translate. In: Proceedings of the International Conference on Learning Representations (ICLR) (2015)
5. Bishop, C.M.: Pattern Recognition and Machine Learning. Springer, New York (2006). https://www.microsoft.com/en-us/research/people/cmbishop/#!prml-book
6. Borenstein, E., Ullman, S.: Class-specific, top-down segmentation. In: Heyden, A., Sparr, G., Nielsen, M., Johansen, P. (eds.) ECCV 2002. LNCS, vol. 2351, pp. 109–122. Springer, Heidelberg (2002). https://doi.org/10.1007/3-540-47967-8_8
7. Bottou, L., Le Cun, Y., Bengio, Y.: Global training of document processing systems using graph transformer networks. In: Proceedings of Conference on Computer Vision and Pattern Recognition (CVPR) (1997)
8. Chandra, S., Kokkinos, I.: Fast, exact and multi-scale inference for semantic image segmentation with deep Gaussian CRFs. In: Leibe, B., Matas, J., Sebe, N., Welling, M. (eds.) ECCV 2016. LNCS, vol. 9911, pp. 402–418. Springer, Cham (2016). https://doi.org/10.1007/978-3-319-46478-7_25
9. Chandra, S., Usunier, N., Kokkinos, I.: Dense and low-rank Gaussian CRFs using deep embeddings. In: Proceedings of International Conference on Computer Vision (ICCV) (2017)

10. Chen, L., Schwing, A., Yuille, A., Urtasun, R.: Learning deep structured models. In: Proceedings of the International Conference on Machine Learning (ICML) (2015)
11. Chen, T.Q., Rubanova, Y., Bettencourt, J., Duvenaud, D.K.: Neural ordinary differential equations. In: Advances in Neural Information Processing Systems 31 (NeuIPS) (2018)
12. Collobert, R., Weston, J., Bottou, L., Karlen, M., Kavukcuoglu, K., Kuksa, P.: Natural language processing (almost) from scratch. J. Mach. Learn. Res. **12**, 2493–2537 (2011)
13. Devlin, J., Chang, M.W., Lee, K., Toutanova, K.: BERT: pre-training of deep bidirectional transformers for language understanding (2018). arXiv:1810.04805
14. Girshick, R., Donahue, J., Darrell, T., Malik, J.: Rich feature hierarchies for accurate object detection and semantic segmentation. In: Proceedings of the IEEE Conference on Computer Vision and Pattern Recognition (CVPR), pp. 580–587 (2014)
15. Goodfellow, I., Bengio, Y., Courville, A.: Deep Learning. MIT Press, Cambridge (2016). http://www.deeplearningbook.org
16. He, K., Zhang, X., Ren, S., Sun, J.: Deep residual learning for image recognition. In: Proceedings of the IEEE Conference on Computer Vision and Pattern Recognition (CVPR) (2016)
17. Hinton, G., et al.: Deep neural networks for acoustic modeling in speech recognition. IEEE Sig. Process. Mag. **29**, 82–97 (2012)
18. Ionescu, C., Vantzos, O., Sminchisescu, C.: Matrix backpropagation for deep networks with structured layers. In: Proceedings of the IEEE International Conference on Computer Vision (ICCV) (2015)
19. Jaderberg, M., Simonyan, K., Vedaldi, A., Zisserman, A.: Deep structured output learning for unconstrained text recognition (2014). arXiv:1412.5903v5
20. Krizhevsky, A., Sutskever, I., Hinton, G.E.: ImageNet classification with deep convolutional neural networks. In: Advances in Neural Information Processing Systems 25 (NeuIPS) (2012)
21. Lafferty, J., McCallum, A., Pereira, F.: Conditional random fields: probabilistic models for segmenting and labeling sequence data. In: Proceedings of the International Conference on Machine Learning (ICML) (2001)
22. Le Cun, Y., Bottou, L., Bengio, Y., Haffner, P.: Gradient based learning applied to document recognition. Proc. IEEE **86**(11), 2278–2324 (1998)
23. Leblond, R., Alayrac, J.B., Osokin, A., Lacoste-Julien, S.: SEARNN: training RNNs with global-local losses. In: Proceedings of International Conference on Learning Representations (ICLR) (2018)
24. LeCun, Y., Chopra, S., Hadsell, R., Ranzato, M., Huang, F.J.: Energy-based models. In: Bakir, G.H., Hofmann, T., Schölkopf, B., Smola, A.J., Taskar, B., Vishwanathan, S.V.N. (eds.) Predicting Structured Data (2006)
25. Lefkimmiatis, S.: Non-local color image denoising with convolutional neural networks. In: Proceedings of the IEEE Conference on Computer Vision and Pattern Recognition (CVPR) (2017)
26. Long, J., Shelhamer, E., Darrell, T.: Fully convolutional networks for semantic segmentation. In: Proceedings of the IEEE Conference on Computer Vision and Pattern Recognition (CVPR) (2015)
27. Maclaurin, D., Duvenaud, D., Adams, R.: Gradient-based hyperparameter optimization through reversible learning. In: Proceedings of the International Conference on Machine Learning (ICML) (2015)

28. Metz, L., Poole, B., Pfau, D., Sohl-Dickstein, J.: Unrolled generative adversarial networks. In: Proceedings of the International Conference on Learning Representations (ICLR) (2017)
29. Osokin, A.: How to put algorithms into neural networks. In: CEUR Workshop Proceedings: Proceedings of the International Conference on Data Analytics and Management in Data Intensive Domains (DAMDID/RCDL) (2019)
30. Paszke, A., Gross, S., Massa, F., et al.: PyTorch: an imperative style, high-performance deep learning library. In: Advances in Neural Information Processing Systems 32 (NeuIPS) (2019)
31. Pérez-Cruz, F., Ghahramani, Z., Pontil, M.: Conditional graphical models. In: Bakir, G.H., Hofmann, T., Schölkopf, B., Smola, A.J., Taskar, B., Vishwanathan, S.V.N. (eds.) Predicting Structured Data. MIT Press, Cambridge (2007)
32. Ren, S., He, K., Girshick, R., Sun, J.: Faster R-CNN: towards real-time object detection with region proposal networks. IEEE Trans. Pattern Anal. Mach. Intell. (TPAMI) **39**(6), 1137–1149 (2017)
33. Ronneberger, O., Fischer, P., Brox, T.: U-Net: convolutional networks for biomedical image segmentation. In: Proceedings of the International Conference on Medical Image Computing and Computer-Assisted Intervention (2015)
34. Shevchenko, A., Osokin, A.: Scaling matters in deep structured-prediction models (2019). arXiv:1902.11088
35. Taskar, B., Guestrin, C., Koller, D.: Max-margin Markov networks. In: Proceedings of Neural Information Processing Systems Conference (NIPS) (2003)
36. Tsochantaridis, I., Joachims, T., Hofmann, T., Altun, Y.: Large margin methods for structured and interdependent output variables. J. Mach. Learn. Res. **6**, 1453–1484 (2005)
37. van den Oord, A., Li, Y., Babuschkin, I., Simonyan, K., et al.: Parallel WaveNet: fast high-fidelity speech synthesis. In: Proceedings of the International Conference on Machine Learning (ICML) (2018)
38. Vu, T., Osokin, A., Laptev, I.: Context-aware CNNs for person head detection. In: Proceedings of International Conference on Computer Vision (ICCV) (2015)
39. Wu, Y., et al.: Google's neural machine translation system: bridging the gap between human and machine translation (2016). arXiv:1609.08144
40. Zheng, S., et al.: Conditional random fields as recurrent neural networks. In: Proceedings of International Conference on Computer Vision (ICCV) (2015)

Methodology for Automated Identifying Food Export Potential

Dmitry Devyatkin[1](✉) and Yulia Otmakhova[2,3]

[1] Federal Research Centre "Computer Science and Control" RAS, Moscow, Russia
devyatkin@isa.ru
[2] Moscow State University, Moscow, Russia
otmakhovajs@yandex.ru
[3] Central Economics and Mathematics Institute RAS, Moscow, Russia

Abstract. The food and agriculture could be a driver of the economy in Russia if intensive growth factors were mainly used. In particular, it is necessary to adjust the food export structure to fit reality better. This problem implies long-term forecasting of the commodity combinations and export directions which could provide a persistent export gain in the future. Unfortunately, the existing solutions for food market forecasting tackle mainly with short-term prediction, whereas structural changes in a whole branch of an economy can last during years. Long-term food market forecasting is a tricky one because food markets are quite unstable and export values depend on a variety of different features.

The paper provides a methodology which includes a multi-step data-driven framework for export gain forecasting and an approach for automated evaluation of related technologies. The data-driven framework itself uses multimodal data from various databases to detect these commodities and export directions. We propose a quantile nonlinear autoregressive exogenous model together with pre-filtering to tackle with such long-term prediction tasks. The framework also considers textual information from mass-media to assess political risks related to prospective export directions. The experiments show that the proposed framework provides more accurate predictions then widely used ARIMA model. The expert validation of the obtained result confirms that the methodology could be useful for export diversification.

Keywords: Data-driven market forecasting · International trade · Quantile regression · Multimodal data

1 Introduction

Sanctions and trade confrontations set difficulties for persistent economic growth. The natural way to overcome them is making the economy more independent and diversified [1]. Due to limited resources, the efforts should be focused on a limited set of development directions. Therefore the developing of a particular economy field implies discovering a restricted set of the new perspective commodity items and export directions. In the same time, large-scale application of digital technologies is underway in

© Springer Nature Switzerland AG 2020
A. Elizarov et al. (Eds.): DAMDID/RCDL 2019, CCIS 1223, pp. 13–28, 2020.
https://doi.org/10.1007/978-3-030-51913-1_2

the world economy; companies are switching to new business development models, hence introducing modern technologies to manage the markets is the essential goal for governments.

In this paper, we consider these problems in the case of food and agriculture field. Our aim consists in finding the pairs *<Trade partner, Agriculture_OR_FoodCommodity>* with a high probability of the persistent export value growth from a particular country (in our case – from Russia) in several next years. More precisely, we predict summary export value gain in the following two years based on information about current and two past years. This goal is not trivial because of the following issues:

1. Unstable character of many trade flows.
2. Too many features influence on trade flows. If we used them all, it would lead to over-complex prediction models, which aren't trainable with the dataset. Long-term forecasting requires the using of complex models that consider a large number of features and parameters, but the size of the training dataset is strictly limited. Therefore, complex models can be easily overfitted and in some cases give incorrect results on unseen data.
3. Political decisions, economic sanctions strongly affect trade flows, but they hardly ever can be predicted using only statistic databases.
4. Existing regression and classification metrics such as MSE or F1-score poorly reflect the accuracy of the solution to the highlighted problem, since even a small ranking error can lead to the omission of a very profitable direction.

In this paper, we propose a data-driven framework which can mitigate the highlighted problems.

At first, we apply a quantile regression loss since it allows estimating the distribution parameter for the predicted value so that we can process unstable trade flows more accurately.

Secondly, we believe it is possible to mitigate the overfitting problem and instability problems both if one pre-filters pairs with high probabilities of a decline in the future. This can be done with training a binary classifier, which is much simpler than regression and can be performed using simpler models which are not overfitted. Then the "large" errors of the regression model will have less impact on the final result. We also propose compositional features which can describe the market demand for a commodity item compactly to simplify the regression model.

Thirdly, we extract sentiment features from texts, more precisely, from news to assess political risks.

Finally, we calculate ratios between the export value of the top predicted pairs and the export value of the actual top pairs with the highest export gain to assess the usefulness of the prediction.

We also suggest an approach for detailed analysis of the obtained commodity items and finding technologies which could help to push the export for these commodities up.

This paper extends [2] with a review of state-of-the-art methods for agent-based export modelling (Sect. 2), with an approach for automated evaluation of technology background (Sect. 3), and with results of export gain forecasting for 2019–2020 as well as with their detailed description.

The rest of the paper is organized as follows: in Sect. 2 we review related studies; in Sects. 3 we present the data-driven framework; Sect. 4 contains the approach for automated evaluation of technologies, in Sect. 5 we describe the results of the experimental evaluation; Sect. 6 contains conclusion and directions of the future work.

2 Related Work

The vastest branch of studies is devoted to short-term food market forecasting with basic regression and autoregression models. For example, Mor with colleagues propose linear regression and Holt-Winters' models to predict short-term demand for dairy products [3]. The more complex autoregressive integrated moving average model (ARIMA) allows dealing with non-stationarity time series. This model also widely used for food market forecasting, for example, in [4] to forecast harvest prices based on past monthly modal prices of maize in particular states.

Ahumada et al. [5] proposed an equilibrium correction model for corn price. Firstly they use an independent model for each corn. Then they also observe whether the forecasting precision of individual price models can be improved by considering their cross-dependence. The results show that prediction quality can be improved using models that include price interactions. The multi-step approach is proposed in [6]. The researchers consider the balance between production and market capacity to be the key factor for trade flows forecasting.

For forecasting in volatile markets, it is necessary to reveal detail information about the distribution of the predicted variables, not their mean values only. Quantile regression is a common solution in this case [7, 8]. For example, researchers [9] apply linear quantile loss to train Support Vector Regressor and use it to assess confidence intervals for predicted values. The paper [10] combines hybrid ARIMA and Quantile Regression (ARIMA-QR) approaches to construct high and low quantile predictions for non-stationary data. The obtained results show that the model yield better forecasts at out-sample data compared to baseline forecasting models.

Let us briefly highlight some studies related to features for food market forecasting which can better explain trade flow dynamic than trade and production values themselves. Paper [11] provides a conclusion that finance matters for export performance, as commodities with higher export-related financial needs disproportionately benefit from better economic development. Jaud with colleagues uses level of outstanding short-term credit and trade credit insurance, reported in the Global Development Finance and Getting Credit Index (EGCc) from the World Bank Doing Business Survey as features related to the level of financial development.

Political factors also influence on the food market. Makombe with colleagues studies the relationship between export bans and food market [12]. The researchers conclude that the prohibitions cause market uncertainty which may have long-run implications for future food security and trade flows. The critical problem here lies in uncertainty in the way how to formalize and consider these factors in models. It is well-known fact, that political decisions often follow by outbursts in mass media, so one can easily predict possible political decisions if he or she analyses the new sentiment. This idea is widely used for short-term analysis in financial markets [13], thus we believe it could be helpful in the proposed framework.

Long-term export prediction assumes considering arbitrary dependencies between the model outcome and the lots of factors in the past. Duration of these dependencies can vary from single days for price movements to dozens of years for political decisions or climate changing. The mentioned approaches cannot model linear, non-linear dependencies and consider a broad set of sophisticated features at the same time though. A natural way to model such complex features and dependencies is to use neural network framework. Pannakkong with colleagues [14] uses a dense multilayer feed-forward network and ARIMA to forecast cassava starch export value. The results show that feed-forward neural network models overcome the ARIMA models in all datasets. Hence, the neural network models can predict the cassava starch exports with higher accuracy than the baseline statistical forecasting method such as the ARIMA. However such a simple architecture can not model long-term interaction.

There are particular network architectures for long-term prediction. In [15], researchers suggested the nonlinear autoregressive exogenous model (NARX) artificial network architecture for market forecasting. They proposed a feed-forward Time Delay Neural Network, i.e. the network without the feedback loop of delayed outputs, which could reduce its predictive performance. The main benefit of the model compared to model compositions is the ability of joint training of linear and non-linear parts of models. Similarly, in [16] authors proved that the generalized regression neural network with fruit fly optimization algorithm (FOA) is effective for forecasting of the non-linear processes.

Unfortunately, neural network approach often leads to inadequately complex models which are needed large datasets to be reliably fitted. We have relatively small dataset thus it is required to find the most straightforward network architecture and tightest feature set which however could achieve satisfactory forecast accuracy.

Because food market is volatile, it would be helpful if the forecasting model provided more information about predicted variables as quantile regression does. Although there are few works in which quantile regression-like loss function was used for training neural networks [17].

The methodological aspects of creating models for export forecasting require further study. Existing models consider some important indicators, but they can be based on erroneous assumptions that cast doubt on the obtained results. For example, the predictive model for assessing the country's diversification of exports provided in the Atlas of Economic Complexity (Feasibility charts) [18]. This model predicts a very curious output, namely that tropical palm oil could be one of the products for diversifying Russia's exports. This is due to the neglecting country's climatic and infrastructure capabilities. That is why the feature set is still not obvious for this problem.

The review shows that the most applicable solution for the food export gain forecasting is to combine long-term prediction models, such as NARX and quantile loss functions. In addition to basic features such as trade flows and production levels, these models should consider heterogeneous macroeconomic and climate indicators. Since the addition of political factors would complicate the regression model, it makes sense to consider them separately. That is, after the regression, we filter obtained export directions if they are related to high political risks.

Supposing the promising commodities and export directions were obtained, it would be useful to assess the applicability of related technologies and activities for enforcing the export. It appears the agent-based approach is the best one for this task. This approach is universal, and it is convenient for applied science researchers and practitioners due to visualization; at the same time, however, it sets high requirements to computational resources [19]. That approach can be used for modelling a spatial structure of the Russian economic system with an inter-regional cross-sectoral "costs-output" model. A project for developing a model of the European economy – EURACE (Agent-Based Computational Economics) is especially exciting, as numerous autonomous agents operate in a social-and-economic system [20]. The agro-food market has been in the focus of agent-modelling studies; in the past several years, agent-based modelling has been used increasingly broader to determine agricultural and industrial policies. For example, Abdou with colleagues proposes models for the European agro-food market [21]. However, there are no recent works on the opportunities for applying the agent approach to modelling the grain market.

3   Framework for Export Gain Forecasting

As a test dataset for the framework, we use annual information about trade flows (from UN Comtrade [22]), production values (UN FAOSTAT [23]) and macroeconomic indicators (International Monetary Foundation [24]). We consider the following macroeconomic features: state GDP, inflation level, population level etc. Due to heuristic reasons the dataset includes only the items which are produced in Russia and presented in its trade flows, so the final dataset contains 70 export directions and 50 commodities. We also do not consider records earlier than 2009, because the international financial crisis could lead to changes, which we cannot model adequately. Daily climate (temperature, wind speed, humidity, pressure) features were downloaded from RP-5 weather database [25]. The highest, the lowest and average values for each season were calculated, because the time step of the framework is one year. We also used open-available Russian news corpus from Kaggle [26].

It is no doubt to say that trade flows between particular country and its partners depends on trade flows between these partners and the other countries. Unfortunately, if we added all these features directly to the regression model, the model would become too complex and would tend to overfit in the dataset. We propose the SPR (Substantial PRoduct) composite features to resolve this problem. The SPR shows contribution of an arbitrary exported commodity item from Russia on the global demand satisfying (expression 1).

$$SPR_i = \frac{X_i}{\sum_{i \in I, j \in D} C_{ij}}, \tag{1}$$

Here I is a set of leading export commodities, D is a set of export directions, X_i – total Russian export value for commodity i, C_{ij} – export value from Russia to country j for commodity i. We consider all trade flows between Russia and its' partners directly and encode the rest flows with the SPR features. The comprehensive feature list is presented in Table 1.

Table 1. Feature set for export gain forecasting

Group of features	Frequency	For	Features
Trade flows	Annual	Country, Commodity	Export value
			Import value
			Re-export value
			Re-import value
SPR	Annual	Commodity	SPR
Production	Annual	Country, Commodity	Production value
Macro-economic indicators	Annual	Country	Trade balance
			GDP
			Inflation (CPI)
			Inflation (PPI)
			Population
			Purchasing power parity (PPP)
			Unemployment rate
Climate indicators	Per season: max, min, median	City, town	Temperature
			Humidity
			Wind speed
			Precipitation
			Pressure
			Cloudiness

We propose the following framework for prediction of the promising pairs <*export item, direction*> (Fig. 1). The framework contains regression step and several filtering steps. Pre- and post-filtering steps are proposed to deal with the trade flows instability.

At first step of the framework we detect pairs which likely tend to decrease. On the one hand the filtering model should be much simpler, than the regression model, but on the other hand it should learn complex non-linear dependencies. The most appropriate approach in this case is decision tree ensembles. We tested several methods such as Random Forest [27], Gradient Boosting [28] and XGBoost [29] to fit these ensembles. The next two steps we realize with the modified NARX quantile regression model (Fig. 2). A single model is used for all directions and commodities since the use of individual models can lead to the loss of information about the interaction between the export value for commodities. We used the following loss function instead of mean squared error to obtain quantile NARX model:

$$L(\omega, \theta) = \sum_{t \vee y_t \geq f(x_t, \omega)} \theta(y_t - f(x_t, \omega)) + \sum_{t \vee y_t < f(x_t, \omega)} (1 - \theta)(y_t - f(x_t, \omega)),$$

$$(2)$$

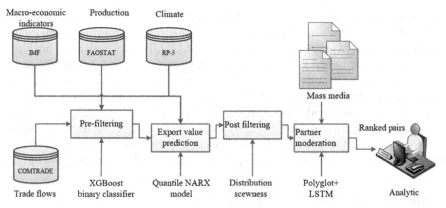

Fig. 1. Framework for prospective export pairs prediction

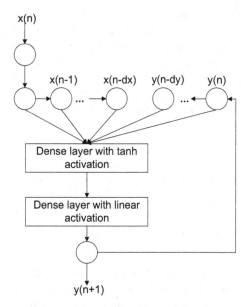

Fig. 2. Nonlinear autoregressive exogenous model for export value forecasting

here θ – quantile level, x_t – features for time t, $f(x_t, \omega)$ – network output for time t and ω – parameters of the network.

This function was firstly introduced in [17]; it is a direct application of quantile regression [7] for networks training. Thanks to error-backpropagation framework the network architecture does not have any affection on the function (2).

The modified NARX model allows predicting values for different quantile levels. We predict export flows with quantile levels 0.25, 05 and 0.75 and assessed skewness of the results. Than pairs with positive distribution skew are filtered. We also applied the Autoregressive Integrated Moving Average (ARIMA) model as a baseline.

In the last step, we filter unreliable trade partners with various models for sentiment analysis. We tested two neural network models, namely Attention-based Long-Short Term Memory (LSTM) [30] and Contextual LSTM [31]. Polyglot sentiment analyzer was also tested as a baseline. We used the Kaggle corpus with more than 10K news reports in Russian to train these models. Post-filtering itself consists of two parts. At first, we apply the Polyglot library [32] together with country name dictionary to extract news, mentioned Russia and some other country together. Then we apply a sentiment analyzer and filter trade partners with highly negative sentiment scores from the results.

The next essential step is a detailed analysis of the obtained commodity items and finding technologies which could help to push the export for these commodities up.

4   Automated Evaluation of Technologies

The initial step of the approach includes defining a set of commodities and ranking them with the complexity index [18]. Next step is filtering the products which Russia does not produce (or can produce), as well as assessing the similarity to other commodities with the higher complexity index. This way, the commodity can fall into another category of complexity if one uses more advanced technologies for primary or deep processing. But the very possibility of that depends on technological background and availability of centers of competence. Of course, determining the possibilities of launching those products requires an analysis of the domestic and foreign market in the short and long term (Fig. 3).

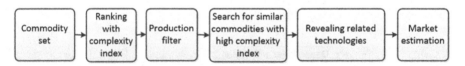

Fig. 3. Essential steps of a technology evaluation

The detection of competency centers presumes gathering and analysis documents related to the technologies. We suggest Scopus for the scientometric analysis because it contains the complete set of scientific publications for agriculture and food research. For patent analysis, we apply LexisNexis [33], and IAS "Priorities" [34], namely LexisNexis is used to search for the world patents, whereas "Priorities" is used to analyze Russian and US patents. The key feature of the approach is that the query is generated automatically with the "Priorities" based on a small subset of technology-related documents [35].

It appears that employing capabilities of agent-based modelling in a complicated macro-economic situation is not only useful for designing new scenarios of developing the agro-food market but is also helpful in assessing the efficiency of measures undertaken by the state to enhance food security with modern digital platform and technologies.

With the turbulence in global trade and the necessity to develop non-commodity exports in Russia, the choice of an effective export strategy in the food market is becoming vitally important these days. Agent-based models (ABM) simulate the behavior of decentralized self-learning agents with their own goals and capabilities [36]. ABM can be used as a tool for analyzing and predicting market movements. According to this approach, an economic system can be represented as a set of interactive subsystem-agents. Thus, it is possible to study the behavior of a food market with changing the features of particular agents and defining the parameters of their interaction..

We plan to change such factors as the level of global demand, the amount of customs duties, the exporting companies' funds in order to determine the strategic conduct of the exporting agents taking into account the limited rationality of the participants in communicative interaction.

5   Results and Discussion

We evaluated the proposed framework as well as its crucial parts. At first, the classification performance for the filtering step was evaluated. We tested several decision tree ensembles (Random Forest, Gradient Boosting, XGBoost) and Linear Support Vector Machine (SVM) classifier as a baseline. The XGBoost method revealed the best accuracy in cross-validation, so we add it to the framework. It is worth to note that the filtering itself can be done with relatively high quality (about 73% F_1 on 5-fold cross-validation, see Table 2, "filtering" column) using a simple feature-set because this task is much simpler than the whole regression task.

Table 2. Evaluation results for filtering methods

Method for filtering	F_1-binary
Random Forest	0.64 ± 0.04
Gradient Boosting	0.59 ± 0.02
XGBoost	0.73 ± 0.05
LinearSVM	0.63 ± 0.02

We also assessed the importance of different types of features for the filtering. The classifier was trained and tested on modified feature sets, in which distinct group of features had been omitted (see Table 3, "filtering" column). This column contains difference between binary F_1 score obtained on the full feature set and the score obtained on clipped feature set. The higher the F_1 score drop is, the more important related subset of features is. Results show that the most important features are SPR, climate and macro-economic indicators.

Then we studied the importance of the different types of features for the regression. As for filtering step, we separated features into distinct subsets and trained the regressor with them (see Table 3, "regression" column). The "Predicted export value gain" here and in the next tables means ratios between the export value of the top-10 predicted pairs and export value of the top-10 actual pairs with the highest export gain. The "ΔPredicted export value gain" column contains difference between the gain obtained on the full feature set and the gain obtained on clipped feature set. The results showed that the most

Table 3. Importance of the particular feature groups for the pre-filtering and for regression results

Group of features	Filtering	Regression
	ΔF_1	ΔPredicted export value gain, in %
Trade flows	0.05	7.1
SPR	0.17	6.8
Production	0.12	1.2
Macro-economic indicators	0.29	24.7
Climate indicators	0.05	17.5

significant features are macro-economic indicators, climate and past export flows. This confirms the limitation of models which do not consider these features, for example [18].

We also evaluated the contribution of the filtering steps to the results. The filtering steps together help significantly improve the obtained results, as one can conclude from Table 4. These steps allow removing the pairs with the highest decline risk.

Table 4. Importance of the particular filtering steps

Filtering	Predicted export value gain, %
Without	12.5
Pre-filtering (only)	38.9
Post-filtering (only)	22.7
All (combined)	61.6

Table 5 contains results for considered sentiment analysis methods. We evaluated these methods on the test subset of the Kaggle sentiment dataset with cross-validation. Attention LSTM model shows slightly better result on this task, so we added it to the proposed framework.

Table 5. Results for the sentiment analysis

Method	F_1-macro
Polyglot sentiment	0.62 ± 0.02
CLSTM	0.79 ± 0.03
Attention LSTM	0.82 ± 0.03

Figure 3 shows average sentiment for mass-media news related to main trade partners of Russia, which are often mentioned in Russian news. The news was gathered from

Lenta.ru dataset [37], because the timeline of the Kaggle dataset is not appropriate. We filtered news, contained both "Russia" and other country names and evaluated average sentiment for them with the Attention LSTM model. It's easy to note that unreliable partners get lower marks (Fig. 4).

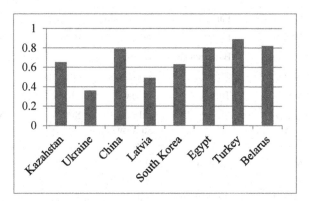

Fig. 4. Partner assessment with sentiment analysis

We tested the overall framework on retrospective data, more precisely records from 2009 to 2014 were used to train and other data (2015–2016) we left for the evaluation. Hyper-parameters for the framework were obtained on train subset with a cross-validation. The detailed results for the whole framework are presented in Table 6. The "Actual" column contains ranked pairs with the highest average export gain in 2015–2016 for Russian Federation. Summary average gain for the top 10 pairs amounted to 1.5 billion USD. The "Predicted" columns contain results of the forecasting.

The economic analysis of the detailed results showed that the list of the partners with the highest summary export gain did not totally match and could extend the list of top importers for the study period.

The proposed NARX model allows predicting the most growing export commodities quite precisely. Linseed is the only mismatched position, but it reflects a new prospective market. Moreover, linseed production for export has been supported by the Russian government since 2016. Thereby the model detected this potential market with past data and predicted that decision.

The most often cause of the NARX model errors is neglecting features, related to technological development. However, the appearance of new technologies leads to dramatic changes in the markets. New deep-processed commodities appear, and prices for existing raw products can decline, which leads to an export value drop for traditional providers. Existing counterparties (Turkey, for example) may switch to other commodities. Therefore there is a need to add technological features to the model, which would make it possible to predict prospect commodities with an assessment of the related technologies for primary and deep processing.

To sum up, the results of the proposed framework could be useful for export diversification since NARX model provides new prospective commodity items. The variants of the NARX model and ARIMA are also helpful for counterparty countries exploration.

Table 6. Detailed results of the export forecasting evaluation

Actual		Predicted			
		ARIMA		NARX + pre-filtering + quantile-filtering	
Partner	Commodity	Partner	Commodity	Partner	Commodity
Bangladesh	Wheat	Egypt	Wheat	Egypt	Wheat
Egypt	Wheat	Bangladesh	Wheat	Saudi Arabia	Barley
China	Soybeans	China	Soybeans	Nigeria	Wheat
Nigeria	Wheat	Turkey	Wheat	Morocco	Wheat
China	Oil of Sunflower Seed	China	Oil of Sunflower Seed	Sudan	Maize
Rep. of Korea	Maize	Algeria	Oil of Soybeans	Turkey	Barley
Lebanon	Wheat	Azerbaijan	Wheat	Turkey	Linseed
Algeria	Oil of Soybeans	Saudi Arabia	Barley	Bangladesh	Wheat
Saudi Arabia	Barley	Lebanon	Wheat	Italy	Wheat
China	Oil of Soybeans	China	Oil of Soybeans	China	Oil of Soybeans
Export value gain, M USD	1474.4; 100%	674.8; 45.7%		908.65; 61.6%	

The proposed framework was also used for export gain forecasting on 2019–2020. We gathered trade flows, weather and macro-economic indicators from the mentioned databases and applied ARIMA to predict production levels for those dates. Table 7 contains the results of the forecasting.

An analysis of the results shows that there are top trade partners with either long-term economic relations (Belarus, Kyrgyzstan, Bulgaria, Morocco, Egypt) or recent promising trade agreements (Oman, Vietnam, Turkey, China). Moreover, all the commodities proposed by the framework fully correspond to the actual positions of modern food imports of the selected partners. Thus, according to the identified pair "China-baby food", it should be noted that China ranks first in the world among im-porters of baby food and is persistently expanding the list of suppliers. The market for baby food in China is very dynamic and shows steady growth, so one needs to pay attention to this promising position for the expansion of Russian exports. The pair "China-rapeseed" reflects the trend of increasing rapeseed export in Russia in 2018. Russia has already become the second-largest supplier to China, and in turn, China receives less dependence on the leading rapeseed exporter, Canada.

The pair "Oman-Sunflower Oil" on a one-hand reflects the trend of diversification of Oman exporters (Ukraine) since 2017 and on the other hand, shows an increase in

Table 7. Detailed results of the export gain prediction

Partner	Commodity	Predicted export gain 2019–2020, M$ (to 2017–2018)
Viet Nam	**Wheat**	1101.8
Turkey	**Wheat**	1031.6
Egypt	**Wheat**	160.3
China	Rapeseed or colza seed	75.1
Morocco	Oil of Soybeans	48.0
Viet Nam	Chicken meat	25.2
Viet Nam	Offals and liver of chickens	24.9
Belarus	**Wheat**	12.6
Kyrgyzstan	Meat, Canned (Poultry)	8.6
Oman	Oil of Sunflower Seed	3.1
Bulgaria	**Wheat**	3.1
China	Infant Food	2.5
Belarus	Cake of Soybeans	2.0
Viet Nam	Sunflower seed	1.0

the export of sunflower oil from Russia. Of interest are the pairs "Vietnam-wheat", "Vietnam-chicken meat", "Vietnam-chicken offal", "Vietnam-sunflower seed", which reflect the recent trade agreements between Vietnam and Russia. For the remaining pairs, it is worth to note that for the selected commodities there is the possibility of increasing the export from Russia to those countries.

Based on the forecast results, one of the most promising directions for increasing export gain is wheat processing technologies. Therefore, we evaluated technological background for those technologies with the proposed approach (On the United States, Russian and World patent databases). The results are presented in Table 8.

The ratio of Russian patents to the world patents for the analyzed technologies reflects the technological background. The table shows that the most promising technologies in terms of existing background are related to the production of particular amino acids as well as food supplements. We also analyzed the dynamics of patent activity and concluded that for most areas the number of patents is growing significantly.

Table 8. The results of the analysis of patent activity in the field of wheat processing (from 2000 to 2018)

Patent database	Products of wheat processing										
	Gluten	Starch	Sorbitol	Glucose	Lysine	Methionine	Threonine	Tryptophan	Glucose-fructose syrup	Bioethanol	Food supplements
Russia	80	391	3	76	77	41	30	30	0	4	25
US	464	1143	20	82	20	24	16	20	0	5	4
World	6762	31852	365	1028	331	337	213	224	11	89	97
Russia/ World	1%	1%	1%	1%	23%	12%	14%	13%	0%	4%	25%

6   Conclusion

In this paper, we propose a methodology which includes a multi-step data-driven framework for export gain forecasting and an approach for automated evaluation of related technologies. The framework considers multimodal open data from many data sources and corpora. In this research, we tried to mitigate the set of problems, related to machine forecasting of food export gain: large feature set dimension, volatility of markets, factors which are difficult to formalize (political risks).

In the experiments, we used open data from FAOSTAT, UN Comtrade, information about global economic situation from International Monetary Foundation, climate information and reports from news corpora. According to the results, quantile loss function and NARX model is a promising combination for long-term prediction of trade flows for food commodities.

The proposed methodology links together the categories of "digital technologies", "demand" and "agri-food industry" by considering many consumer features of the final commodities; therefore this methodology can be considered as a base for a digital platform for identifying technological development priorities for the food market.

In the future research we plan to consider logistical and infrastructure conditions as well as technological features in the framework. We are also going to apply agent modelling to assess the efficiency of different approaches and technologies for pushing the export gain of the obtained commodities.

Acknowledgements. The project is supported by the Russian Foundation for Basic Research, project number 18-29-03086. The project is also partially funded by the project "Text mining tools for big data" as a part of the program supporting Technical Leadership Centers of the National Technological Initiative "Center for Big Data Storage and Processing" at the Moscow State University (Agreement with Fund supporting the NTI projects No. 13/1251/2018 11.12.2018).

References

1. Awokuse, T.: Does agriculture really matter for economic growth in developing countries? In: 2009 Annual Meeting. Agricultural and Applied Economics Association, Milwaukee, Wisconsin, vol. 49762 (2009)
2. Devyatkin, D., Otmakhova, Y.: Framework for automated food export gain forecasting. In: CEUR Workshop Proceedings, vol. 2523, pp. 22–33 (2019)
3. Mor, R.S., Jaiswal, S.K., Singh, S., Bhardwaj, A.: Demand forecasting of the short-lifecycle dairy products. In: Chahal, H., Jyoti, J., Wirtz, J. (eds.) Understanding the Role of Business Analytics, pp. 87–117. Springer, Singapore (2019). https://doi.org/10.1007/978-981-13-133 4-9_6
4. Darekar, A., Reddy, A.: Price forecasting of maize in major states. Maize J. 6(1&2), 1–5 (2017)
5. Ahumada, H., Cornejo, M.: Forecasting food prices: the case of corn, soybeans and wheat. Int. J. Forecast. 32(3), 838–848 (2016)
6. Burlankov, S., Ananiev, M., Gazhur, A., Sedova, N., Ananieva, O.: Forecasting the development of agricultural production in the context of food security. Sci. Pap. Ser.-Manag. Econ. Eng. Agric. Rural Dev. 18(3), 45–51 (2018)
7. Koenker, R., Hallock, K.: Quantile regression. J. Econ. Perspect. 15(4), 143–156 (2001)
8. Maciejowska, K., Nowotarski, J., Weron, R.: Probabilistic forecasting of electricity spot prices using Factor Quantile Regression Averaging. Int. J. Forecast. 32(3), 957–965 (2016)
9. Li, G., Xu, S., Li, Z., Sun, Y., Dong, X.: Using quantile regression approach to analyze price movements of agricultural products in China. J. Integr. Agric. 11(4), 674–683 (2012)
10. Arunraj, N., Ahrens, D.: A hybrid seasonal autoregressive integrated moving average and quantile regression for daily food sales forecasting. Int. J. Prod. Econ. 170, 321–335 (2015)
11. Jaud, M., Kukenova, M., Strieborny, M.: Financial development and sustainable exports: evidence from firm-product data. World Econ. 38(7), 1090–1114 (2015)
12. Makombe, W., Kropp, J.: The effects of Tanzanian maize export bans on producers' welfare and food security. In: Selected Paper Prepared for Presentation at the Agricultural & Applied Economics Association, Boston, MA, vol. 333-2016-14428 (2016)
13. Nassirtoussi, A.K., Aghabozorgi, S., Wah, T.Y., Ngo, D.C.L.: Text mining of news-headlines for FOREX market prediction: a Multi-layer Dimension Reduction Algorithm with semantics and sentiment. Expert Syst. Appl. 42(1), 306–324 (2015)
14. Pannakkong, W., Huynh, V.N., Sriboonchitta, S.: ARIMA versus artificial neural network for Thailand's cassava starch export forecasting. In: Huynh, V.N., Kreinovich, V., Sriboonchitta, S. (eds.) Causal Inference in Econometrics. SCI, vol. 622, pp. 255–277. Springer, Cham (2016). https://doi.org/10.1007/978-3-319-27284-9_16
15. Menezes Jr., J.M.P., Barreto, G.A.: Long-term time series prediction with the NARX network: an empirical evaluation. Neurocomputing 71(16–18), 3335–3343 (2008)
16. Li, H., Guo, S., Sun, J.: A hybrid annual power load forecasting model based on generalized regression neural network with fruit fly optimization algorithm. Knowl.-Based Syst. 37, 378–387 (2013)
17. Taylor, J.W.: A quantile regression neural network approach to estimating the conditional density of multiperiod returns. J. Forecast. 19(4), 299–311 (2000)

18. The atlas of economic complexity. http://atlas.cid.harvard.edu. Accessed 01 July 2019
19. Hamill, L., Gilbert, N.: Agent-Based Modelling in Economics. Wiley, Hoboken (2015). 246 p.
20. Deissenberg, C., van der Hoog, S., Herbert, D.: EURACE: a massively parallel agent-based model of the European economy. Appl. Math. Comput. **204**(2), 541–552 (2008). https://doi.org/10.1016/j.amc.2008.05.116
21. Abdou, M., Hamill, L., Gilbert, N.: Designing and building an agent-based model. In: Heppenstall, A., Crooks, A., See, L., Batty, M. (eds.) Agent-Based Models of Geographical Systems. Springer, Dordrecht (2012). https://doi.org/10.1007/978-90-481-8927-4
22. UN Comtrade: International Trade Statistics. https://comtrade.un.org/data/. Accessed 28 Apr 2019
23. Food and Agriculture Organization of the United Nations. http://www.fao.org/faostat/en/. Accessed 28 Apr 2019
24. International Monetary Foundation. http://www.imf.org/en/Data. Accessed 28 Apr 2019
25. RP5 Weather Archive. http://rp5.ru. Accessed 28 Apr 2019
26. Kaggle Russian news dataset for sentiment analysis. https://www.kaggle.com/c/sentiment-analysis-in-russian/overview. Accessed 28 Apr 2019
27. Breiman, L.: Random forests. Mach. Learn. **45**(1), 5–32 (2001). https://doi.org/10.1023/A:1010933404324
28. Friedman, J.H.: Greedy function approximation: a gradient boosting machine. Ann. Stat. **29**, 1189–1232 (2001)
29. Chen, T., Guestrin, C.: XGBoost: a scalable tree boosting system. In: Proceedings of the 22nd ACM SIGKDD International Conference on Knowledge Discovery and Data Mining, pp. 785–794. ACM (2016)
30. Wang, Y., Huang, M., Zhao, L., Zhu, X.: Attention-based LSTM for aspect-level sentiment classification. In: Proceedings of the 2016 Conference on Empirical Methods in Natural Language Processing, pp. 606–615. Association for Computational Linguistics, Austin (2016)
31. Ghosh, S., Vinyals, O., Strope, B.: Contextual LSTM (CLSTM) models for large scale NLP tasks. In: arXiv preprint arXiv:1602.06291. ACM (2016)
32. Al-Rfou, R., Kulkarni, V., Perozzi, B.: POLYGLOT-NER: massive multilingual named entity recognition. In: Proceedings of the 2015 SIAM International Conference on Data Mining, pp. 586–594. Society for Industrial and Applied Mathematics (2015)
33. Bengisu, M., Ramzi, N.: Forecasting emerging technologies with the aid of science and technology databases. Technol. Forecast. Soc. Change **73**(7), 835–844 (2006)
34. Sokolov, I., Grigoriev, O., Tikhomirov, I., Suvorov, R., Zhebel, V.: On creating a national system for identifying research and development priorities. Sci. Tech. Inf. Process. **46**(1), 14–19 (2019). https://doi.org/10.3103/S0147688219010039
35. Shvets, A., Devyatkin, D., Sochenkov, I., Tikhomirov, I., Popov, K., Yarygin, K.: Detection of current research directions based on full-text clustering. In: 2015 Science and Information Conference (SAI), pp. 483–488. IEEE (2015)
36. Railsback, S.F., Grimm, V.: Agent-Based and Individual-Based Modeling: A Practical Introduction. Princeton University Press, Princeton (2019)
37. Lenta.ru Russian News Dataset. https://github.com/yutkin/Lenta.Ru-News-Dataset. Accessed 28 Apr 2019

Data Infrastructures and Integrated Information Systems

Integration of the JINR Hybrid Computing Resources with the DIRAC Interware for Data Intensive Applications

Vladimir Korenkov[1,3], Igor Pelevanyuk[1,3(✉)], and Andrei Tsaregorodtsev[2,3]

[1] Joint Institute for Nuclear Research, Dubna, Russia
{korenkov,pelevanyuk}@jinr.ru
[2] CPPM, Aix-Marseille University, CNRS/IN2P3, Marseille, France
atsareg@in2p3.fr
[3] Plekhanov Russian Economics University, Moscow, Russia

Abstract. Scientific data and computing-intensive applications become more and more widely used. Different computing solutions have different protocols and architectures, they should be chosen carefully in the design of computing projects of large scientific communities. In a modern world of diverse computing resources such as grids, clouds, and supercomputers the choice can be difficult. Therefore, software developed for integration of various computing and storage resources into a single infrastructure, the so-called interware, makes this choice easier. The DIRAC interware is one of these products. It proved to be an effective solution for many experiments in High Energy Physics and some other areas of science providing means for seamless access to distributed computing and storage resources. The DIRAC interware was deployed in the Joint Institute for Nuclear Research to serve the needs of different scientific groups by providing a single interface to a variety of computing resources: grid cluster, computing cloud, supercomputer Govorun, disk and tape storage systems. The DIRAC based solution was proposed for the currently operational Baryonic Matter at Nuclotron experiment as well as for the future experiment Multi-Purpose Detector at the Nuclotron-based Ion Collider fAcility. Both experiments have requirements making the use of heterogeneous computing resources necessary. A set of tests was introduced in order to demonstrate the performance of the JINR distributed computing system.

Keywords: Grid computing · Hybrid distributed computing systems · Supercomputers · DIRAC

1 Introduction

Data intensive applications became now an essential means for getting insights of new scientific phenomena while analyzing huge data volumes collected by modern experimental setups. For example, the data recording to the tape system at CERN exceeded in total 10 Petabytes per month in 2018 for all the 4 LHC experiments. In 2021, with the start of the Run 3 phase of the LHC program, the experiments will resume data taking

© Springer Nature Switzerland AG 2020
A. Elizarov et al. (Eds.): DAMDID/RCDL 2019, CCIS 1223, pp. 31–46, 2020.
https://doi.org/10.1007/978-3-030-51913-1_3

with considerably increased rates. The needs for the computing and storage capacity of the LHCb experiment, for instance, will increase by an order of magnitude [1].

In a more distant future, with the start of the LHC Run 4 phase, the projected data storage needs of the experiments are estimated to exceed 10 Exabyte's. These are unprecedented volumes of data to be processed by distributed computing systems, which are being adapted now to cope with the new requirements.

Other scientific domains are quickly approaching the same collected data volumes: astronomy, brain research, genomics and proteomics, material science [3]. For example, the SKA large radio astronomy experiment [2] is planned to produce about 3 Petabytes of data daily when it will come into full operation in 2023.

The needs of the LHC experiments in data processing were satisfied by the infrastructure of the World LHC Computing Grid (WLCG). The infrastructure still delivers the major part of computing and storage capacity for these experiments. It is well suited for processing the LHC data ensuring massively parallel data treatment in a High Throughput Computing (HTC) paradigm. WLCG succeeded in putting together hundreds of computing centers of different sizes but with similar properties, typically providing clusters of commodity processors under control of one of the batch systems, e.g. LSF, Torque or HTCondor. However, new data analysis algorithms necessary for the upcoming data challenges require a new level of parallelism and new types of computing resources. These resources are provided, in particular, by supercomputers or High-Performance Computing (HPC) centers. The number of HPC centers is increasing and there is a clear need in setting up infrastructures allowing scientific communities to access multiple HPC centers in a uniform way as it is done in the grid systems.

Another trend in massive computing consists in provisioning resources via cloud interfaces. Both private and commercial clouds are available now to scientific communities. However, the diversity of interfaces and usage policies makes it difficult to use multiple clouds for applications of a particular community. Therefore, providing uniform access to resources of various cloud providers would increase flexibility and the total amount of available computing capacity for a given scientific collaboration.

Large scientific collaborations typically include multiple participating institutes and laboratories. Some of the participants have considerable computing and storage capacity that they can share with the rest of the collaboration. With the grid systems this can be achieved by installing complex software, the so-called grid middleware, and running standard services like Computing and Storage Elements. For managers of local computing resources who are usually not experts in the grid middleware, this represents a huge complication and often results in underused resources that would otherwise be beneficial for the large collaborations. Tools for easy incorporation of such resources can considerably increase efficiency of their usage.

The DIRAC Interware project provides a framework for building distributed computing systems using resources of all different types mentioned above and putting minimal requirements on the software and services that should be operated by the resources providers [4]. Developed originally for the LHCb experiment at LHC, CERN, the DIRAC Interware was generalized to suite requirements of a wide range of applications. It can be used to build independent distributed computing infrastructures as well as to provide services for existing projects. DIRAC is used by a number of High Energy Physics and

Astrophysics experiments but it is also providing services for several general-purpose grid infrastructures, for example, national grids in France [5] and Great Britain [6]. The EGI Workload Manager is the DIRAC service provided as part of the European Grid Infrastructure service catalog. It is one of the services of the European Open Science Cloud (EOSC) project inaugurated at the end of 2018 [7]. The EGI Workload Manager provides access to grid and cloud resources of the EGI infrastructure for more than 500 registered users.

This article is an extension of work originally presented in conference "Data Analytics and Management in Data Intensive Domains" (DAMDID/RCDL'2019) [4]. In this paper we give in Sect. 2 an overview of the DIRAC Interware, its capabilities and further development directions. In Sect. 3 computing facilities at the Joint Institute for Nuclear Research (JINR), Dubna are presented. We describe the DIRAC based infrastructure deployed at JINR putting together a number of local computing clusters as well as connecting cloud resources from JINR member institutions in Sect. 4. The results of storage and CPU performance evaluation of the integrated JINR computing infrastructure are presented in Sect. 5.

2 DIRAC Interware

The DIRAC Interware project provides a development framework and a large number of ready-to-use components to build distributed computing systems of arbitrary complexity. DIRAC services ensure integration of computing and storage resources of different types and provide all the necessary tools for managing user tasks and data in distributed environments [8]. Managing both workloads and data within the same framework increases the efficiency of data processing systems of large user communities while minimizing the effort for maintenance and operation of the complete infrastructure.

The DIRAC Workload Management System (WMS) architecture is based on the use of multiple autonomous agents which are deployed on computing resources using specific payload submission mechanisms in each case. This can be job submission to computing cluster batch systems, creation of appropriately contextualized virtual machines (VMs) in the cloud services or starting the agents as payloads on the BOINC volunteer computers. In all the cases, the agents inspect their running environment, evaluate the available computing capacity and present the description of the reserved resources to the DIRAC central Task Queue service. On the other hand, users are submitting their tasks to the central Task Queue where they are classified and prioritized according to the community policies. An efficient matching mechanism allows to quickly find the most suitable user task to each agent request taking into account the task requirements and priority. The agent-based WMS increases the efficiency of user payloads execution in a distributed system due to a preliminary validation of the execution environment reducing considerably execution failure rates. The agent-based WMS allows also to abstract away differences in access protocols of heterogeneous computing resources providers and aggregate them as a set of similar logical computing clusters, thus solving the problem of the resource's heterogeneity [9].

Similarly to WMS, the DIRAC Data Management System (DMS) provides an abstract Storage Element interface together with its implementation for the most common modern storage access protocols (SRM, (GRID) FTP, HTTP, WebDAV, S3, etc).

Together with a central File Catalog service, which keeps traces of all the physical file replicas, the DIRAC DMS presents disparate distributed storage resources to the users as a single file system with a unique logical name space.

The DIRAC software is constantly evolving to follow changes in the technology and interfaces of available computing and storage resources. As a result, most existing HTC, HPC and cloud resources can be interconnected with the DIRAC Interware. In order to meet the needs of large scientific communities, the computing systems should fulfill several requirements. In particular, it should be easy to describe, execute and monitor complex workflows in a secure way respecting predefined policies of usage of common resources by multiple user communities.

2.1 Massive Operations

Usual workflows of large scientific collaborations consist in creation and execution of large numbers of similar computational and data management tasks. DIRAC is providing support for massive operations with its Transformation System. The system allows definition of Transformations – recipes to create certain operations triggered by the availability of data with required properties. Operations can be of any type: submission of jobs to computing resources, data replication or removal, etc. Each Transformation consumes some data and derives ("transforms") new data, which, in turn, can form input for another Transformation. Therefore, Transformations can be chained creating data driven workflows of any complexity. Data production pipelines of large scientific communities based on DIRAC are using heavily the Transformation System defining many hundreds of different Transformations. Each large project develops its own system to manage large workflows each consisting of many Transformations. The DIRAC Production System is based on the experience of several community specific workflow management systems and provides a uniform way to create a set of Transformations interconnected via their input/output data filters. It helps production managers to monitor the execution of so created workflows, evaluate the overall progress of the workflow advancement and validate the results with an automated verification of all the elementary tasks.

2.2 Multi-community Services

Multiple relatively small scientific communities have access to common grid and cloud infrastructures but they cannot afford setting up and operating such complex services as DIRAC because of the lack of expertise and manpower. Therefore, services built with the DIRAC Interware can be provided by large infrastructure projects like national grids or JINR distributed cloud. DIRAC services can be configured to support multiple communities with comprehensive rules describing resources access rights for each user group. User communities can register in the DIRAC Configuration System additional resources, e.g. local university computing clusters, and make them available through the same interface as infrastructure resources. The user tasks monitoring and consumed resources accounting is provided per user and per community allowing to efficiently apply common quotas and policies of usage of the common infrastructure.

The security aspects are very important in distributed computing systems with multi-community access. In most of the currently existing multi-community grid infrastructures the security of all operations is based on the X509 PKI infrastructure. In this solution, each user has to, first, obtain a security certificate from one of recognized Certification Authorities (CA). The certificate should be then registered in a service holding a registry of all the users of a given Virtual Organization (VO). The user registry keeps the identity information together with associated rights of a given user. In order to access grid resources, users are generating proxy certificates which can be delegated to grid remote services in order to perform operations on the user's behalf.

The X509 standard based security is well supported in academia institutions but is not well suited for other researchers, for example, working in universities. On the other hand, there are well-established industry standards developed mostly for the web applications that allow identification of users as well as delegation of user rights to remote application servers. Therefore, grid projects started migration to the new security infrastructure based on the OAuth2/OIDC technology. With this technology, user's registration is done by local identity providers, for example, a university LDAP index. On the grid level a Single-Sign-On (SSO) solution is provided by a federation of multiple identity providers to ensure mutual recognition of user security tokens. In particular, the EGI infrastructure has come up with the Check-In SSO service as a federated user identity provider.

The DIRAC user management subsystem was recently updated in order to support this technology. Users can be identified and registered in DIRAC based on their SSO tokens containing also additional user metadata, which, in turn, defines user rights within the DIRAC framework. The DIRAC implementation of the new security framework is generic and can be easily configured to work with multiple identity providers and SSO systems.

2.3 DIRAC Software Evolution

The intensity of usage of the DIRAC services is increasing and the software must evolve to cope with the new requirements. This process is mostly driven by the needs of the LHCb experiment, which remains the main consumer and developer of the DIRAC software. As was mentioned above, the order of magnitude increase in the data acquisition rate of LHCb in 2021 dictates a revision of the technologies used in its data processing solutions.

Several new technologies were introduced recently into the DIRAC software stack. The use of Message Queue (MQ) services allows passing messages between distributed DIRAC components in an asynchronous way with the possibility of message buffering in case of system congestion.

The DIRAC service's states are kept in relational databases using MySQL servers. The MySQL databases have shown very stable operation over the years of usage. However, the increased amount of data to be stored in databases limits the efficiency of queries and new solutions are necessary. The so-called NoSQL databases have excellent scalability properties and can help in increasing the efficiency of the DIRAC components. The ElasticSearch NoSQL (ES) database solution was applied in several DIRAC subsystems relying on heavy queries on loosely structured data, e.g. activities monitoring or service logs inspection.

DIRAC from the moment of its inception used a custom client/service protocol to maximize the efficiency of queries. However, the current implementations of the secure HTTP protocol became very efficient with supporting software packages available, which follow all the evolutions of the protocol versions and security standards. Therefore, the next generation of the DIRAC client/server protocol will be based on the HTTP standard. This can help DIRAC acceptance by new communities with strong security requirements.

These and other additions and improvements in the DIRAC software aim at the overall increase of the system efficiency and scalability to meet requirements of multiple scientific communities relying on DIRAC services for their computing projects.

3   JINR Computing Resources

The Joint Institute for Nuclear Research is an international intergovernmental organization, a world-famous scientific center that is a unique example of the integration of fundamental theoretical and experimental research. It consists of seven laboratories: Laboratory of High Energy Physics, Laboratory of Nuclear Problems, Laboratory of Theoretical Physics, Laboratory of Neutron Physics, Laboratory of Nuclear Reactions, Laboratory of Information Technologies, Laboratory of Radiation Biology. Each laboratory is comparable with a large institute in the scale and scope of the research activity.

JINR has a powerful highly productive computing environment that is integrated into the world computer network through high-speed communication channels. The basis of the computer infrastructure of the Institute is the Multifunctional Information Computer Complex (MICC). It consists of several large components: grid cluster, computing cloud, supercomputer Govorun. Each component has its features, advantages, and disadvantages. Different access procedures, different configuration and connection with different storage systems make it difficult to use all the facilities together for one set of tasks.

3.1   Computing Resources

Grid Cluster. The JINR grid infrastructure is represented by a Tier1 center for the CMS experiment at the LHC and a Tier2 center.

After the recent upgrade, the data processing system at the JINR CMS Tier1 consists of 415 64-bit nodes: 2 x CPU, 6–16 cores/CPU that form altogether 9200 cores for batch processing [10]. The Torque 4.2.10/Maui 3.3.2 software (custom build) is used as a resource manager and a task scheduler. The computing resources of the Tier2 center consist of 4,128 cores. The Tier2 center at JINR provides data processing for all the four experiments at the LHC (Alice, ATLAS, CMS, LHCb) and in addition supports many virtual organizations (VO) that are not members of the LHC (BES, BIOMED, COMPASS, MPD, NOvA, STAR, ILC).

Grid cluster is an example of a n HTC type facility. It means that the primary task of this cluster is to run thousands of independent processes at the same time. Independent means that once a process has started and until it finishes, the process does not rely on any input that is being produced at the same moment by other processes.

Jobs may be sent to the grid using a CREAM Computing Element – a service installed in JINR specially for the grid jobs. Computing element works as an interface to the local batch farm. Its primary task is to authenticate the owner of the job and redirect it to the right queue. Users are required to have an X509 certificate and be a member of a Virtual Organization supported by the Computing Element.

Cloud Infrastructure. The JINR Cloud [11] is based on an open-source platform for managing heterogeneous distributed data center infrastructures - OpenNebula 5.4. The JINR cloud resources were increased up to 1564 CPU cores and 8.1 TB of RAM in total. Cloud infrastructure is used primarily for two purposes: to create personal virtual machines and to create virtual machines to serve as worker nodes for jobs. We are going to focus on the second case.

The biggest advantage of cloud resources as computing capacity is their flexibility. In the case of grid or batch resources, several jobs working on one worker node share between them: operating system, CPU cores, RAM, HDD/SSD storage, disk Input/Output capabilities, and network bandwidth. If a job needs more disk space or RAM it is not straightforward to submit the job to the grid without the help of administrators, who in most cases have to create a dedicated queue for this particular kind of job. In the case of clouds, it is much easier to provide a specific resource suiting the job requirements. It may be a virtual machine with a large disk, specific operating system, required number of CPU cores, RAM capacity and network.

When a job destined to the cloud enters the system the corresponding virtual machine is created by DIRAC using the OpenNebula API. During the contextualization process, the DIRAC Pilot is installed in the VM and configured to receive jobs for this cloud resource. Once the job is finished, the pilot attempts to get the next job. If there are no more jobs for the cloud, the pilot will request the VM shutdown. The pilot in the cloud environment is not limited by the time and may work for weeks. These features make cloud resources perfectly suitable for specific tasks with unusual requirements.

Govorun Supercomputer. The Supercomputer Govorun was put into production in March 2018 [12]. It is a heterogeneous platform exploiting several processors' technologies: a GPU part and a CPU part. The GPU part unites 5 servers DGX-1. Each server hosts 8 NVIDIA Tesla V100 processors. The CPU part is a highly dense liquid-cooled system. The processors inside are Intel® Xeon® Platinum 8268 (80 servers). The total performance of all the three parts is 1 PFlops for operations with single precision and 0.5 PFlops for double precision. SLURM 14.11.6 is used as the local workload manager.

The supercomputer is used for tasks, which require massive parallel computations. For example: to solve problems of lattice quantum chromodynamics for studying the properties of hadronic matter with high energy density and baryon charge and in the presence of strong electromagnetic fields, mathematical modeling of the antiproton-proton and antiproton-nucleus collisions with the use of different generators. It is also used for simulation of the collision dynamics of relativistic heavy ions for the future MPD experiment at the NICA collider.

Right now, the supercomputer utilizes its own authentication and authorization system. Every user of the supercomputer should be registered and allowed to send jobs.

Sometimes, a part of the supercomputer is free from parallel tasks and may be used as a standard batch system. Special user was created for the DIRAC service at JINR. All jobs sent to the Govorun are executed with this user identity. This frees actual users from additional registration procedures delegating user profile management and resource consumption accounting to DIRAC.

3.2 Storage Resources

EOS Storage on Disks. EOS [13] is a multi-protocol disk-only storage system developed at CERN since 2010 for the High Energy Physics experiments (including the LHC experiments). Having a highly-scalable hierarchical namespace, and with the data access possible via the xRootD protocol, it was initially used for the physics data storage. Today, EOS provides storage for both physics experiments and user use cases. For the user authentication, EOS supports Kerberos (for the local access) and X.509 certificates for the grid access. To ease the experiment's workflow integration, SRM as well as GridFTP access is provided. EOS supports the xRootD third-party copy mechanism from/to other xRootD enabled storage services.

The EOS was successfully integrated into the MICC structure. The NICA experiments already use EOS for data storage. At the moment there are ~200 TB of "raw" BM@N data and ~84 GB of simulated MPD data stored in the EOS instance. EOS is visible as a local file system on the MICC worker nodes. It allows users authorized by the Kerberos5 protocol to read and write data. A dedicated service was installed to allow usage of X509 certificates with VOMS extensions.

dCache Disk and Tape Storage. The core part of the dCache has been proven to efficiently combine heterogeneous disk storage systems of the order of several hundred TBs and present its data repository as a single filesystem tree. It takes care of the data integrity, hardware failures and provides that a minimal configured number of copies of each dataset resides within the system to ensure high data availability in the case of disk server maintenance or failure. Furthermore, dCache supports a large set of standard access protocols to the data repository and its namespace. It supports DCAP, SRM, GridFTP, and xRootD protocols [14].

dCache at JINR consists of two parts: disk storage and tape storage. The disk part operations are similar to EOS. The tape storage is accessed through dedicated disk buffer servers. When data is uploaded to the dCache tape part, it is first uploaded to the disk buffer. If the disk buffer is occupied above a certain threshold (which is 80% in our case), all the data is moved from disk to tape and removed from the disk buffer. While the data stays in the buffer, access to it is similar to the access to the dCache disk data. But once the data is moved to the tape and removed from the disk, it will be accessed with a certain delay. The time required to select the right tape and transfer data from the tape to the disk depends on the tape library task queue. Generally, the time varies from 20 s to up to several minutes.

Tape library should be used only for archive storage and preferably for big files. Otherwise, it may incur an unnecessary load on the tape system.

Ceph Storage. Software-defined storage (SDS) based on the Ceph technology is one of the key components of the JINR cloud infrastructure. It has been running in production mode since the end of 2017. It delivers object, block and file storage in one unified system. Currently, the total amount of raw disk space in the SDS is about 1 PB. Due to triple replication, effective disk space available for users is about 330 TB. Users of Ceph can attach part of the storage to a computer using the FUSE disk mounting mechanism. After that, it is possible to read and write data to the remote storage as if it is connected directly to the computer.

The Ceph storage was integrated into the DIRAC installation for tests. Since Ceph does not allow authentication by X509 certificates with VOMS extensions, a dedicated virtual machine was configured to host a DIRAC Storage Element – a standard service which works as a proxy to a file system. It checks certificates with VOMS extensions before allowing writing and reading to a dedicated directory. Right now, Ceph storage does not allow massive transfers since it relies on one server with Ceph attached by FUSE. The test demonstrated that the maximum speed of transfers is not exceeding 100 MB/s which is a consequence of 1 Gb network connection. The way to increase the performance of this storage is an improvement of the network speed up to 10 Gb/s and a possible creation of additional DIRAC Storage Elements which can share the load between them.

4 JINR DIRAC Installation

The DIRAC installation in JINR consists of 4 virtual machines. Three of them are placed on a dedicated server to avoid network and disk I/O interference with other virtual machines. The operating system on these virtual machines is CentOS 7. It appeared that some of the LCG software related to grid job submission is not compatible with CentOS. To cope with that, we created a new virtual machine with Scientific Linux 6 installed there. Flexibility of the DIRAC modular architecture allowed us to do that easily. The characteristics of the virtual machines hosting DIRAC services are presented in Table 1.

Table 1. Virtual machines hosting DIRAC services

	dirac-services	dirac-conf	dirac-web	dirac-sl6
OS	CentOS	CentOS	CentOS	Scientific Linux
Version	7.5	7.5	7.5	6.10
Cores	8	4	4	2
RAM	16 GB	8 GB	8 GB	2 GB

4.1   Use Cases in JINR

Up to now, we foresee two big possible use cases: Monte-Carlo generation for Multi-Purpose Detector (MPD) at NICA and data reconstruction for Baryonic Matter at Nuclotron(BM@N).

Raw data is received by the BM@N detector and uploaded to the EOS storage. There are two data taking runs available now: run 6 and run 7. The data sizes are respectively: 16 TB and 196 TB. All the data is split in files, 800 files for run 6 and around 2200 files for run 7. The main difficulty with these files is the fact that their sizes are very different: from several MBs to up to 800 GBs per one file. This makes data processing a tough task especially on resources with a small amount of local storage or a bad network connection. The data could be processed using the Govorun Supercomputer, but the EOS storage may be accessed only over the xRootD protocol. The data may require full reprocessing one day, if the reconstruction algorithms are changed.

So far, the best solution would be to process big files in the cloud, other files in the grid infrastructure and sometimes, when the supercomputer has free job slots, do some processing there. But without some central Workload Management system and Data Management system this is a difficult task. The data could be placed not only in EOS but also in dCache. This would allow data delivery to the worker nodes using grid protocols like SRM or xRootD. Once the X509 certificates start working for the EOS storage, it will also be included in the infrastructure and be accessible from everywhere.

The second use case is the Monte-Carlo generation for the MPD experiment. The Monte-Carlo generation could be performed almost on all the components of MICC at JINR. It is a CPU intensive task less demanding in terms of disk space and input/output rates. The file size could be tuned to be in a particular range for the convenience of the future storage and use. The use of a central distributed computing system may not be critical right now, but it will definitely be useful later, when the real data arrives. It would allow for design and testing of production workflows, and provide access to different organizations to participate in the analysis of the experiment data.

5   JINR Resource Performance Evaluation

While using hybrid computing resources it is not easy to predict the exact outcome of the job's execution. Too many variables are involved: network speed, disk I/O, different operating systems, different processors. Therefore, it is impossible to plan the load and data flows without additional characterization of the resources.

There are two ways to obtain this information. The first option is to collect details about all the components and estimate the approximate performance. It is a lot of work to collect and analyze this data as well as to keep this information up to date. Any major change or upgrade on one of the resources will make this information obsolete and would require repeated analysis. The results will have to be confirmed by real jobs sent to the system. The advantage of this approach is that it does not affect the resources and their performance itself. It may work on small infrastructures, but on big ones it would require too much administrative work related to data collection and analysis.

The second approach is to prepare a set of functional tests and run it from time to time to collect all important metrics and compare them with each other. This approach may be dangerous since these tests may affect other jobs running on the same shared resources. The tests can interrupt execution of other jobs or data transfers. So, certain care should be taken to run those tests. The advantage of this approach is that it provides real metrics for the resource's performance.

For the purpose of performance evaluation of the JINR resources, the second approach has been chosen since it is accurate, consistent, easy to reproduce. It is possible to coordinate with administrators of local resources and reduce the chance of interruption of any work. Two main measurements were chosen as the most important ones: data download from Storage Elements and CPU performance.

5.1 Performance Test of Storage Resources

In the case of massive data processing, it is crucial to know the limitations of different components. The limitations may depend on the computing and storage resource usage patterns. In most cases, jobs need to download some amount of data before execution of the user application payload. In order to characterize Storage Elements, a synthetic test was proposed: run many jobs on one computing resource, make them start downloading all the same data at the same moment, measure how much time it takes to get the file.

Every test job had to go through the following steps:

1. Start execution on the worker node.
2. Check out the transfer start time.
3. Wait until the transfer starting time.
4. Start the transfer.
5. When the transfer is finished, report the duration of the operation.
6. Remove the downloaded file.

Two storage systems were chosen for the tests: EOS and dCache since only those systems were accessible for both reading and writing on all the computing resources at the moment of the tests. The test file size was chosen to be 3 GB as the most typical one. The amount of the test jobs in one test campaign depended on the number of free CPU cores in our infrastructure. We submitted 200 jobs during one test campaign. Not all of them could start at the same time, which means that during the test less than 200 jobs may download data. This is taken into account when we calculate the total transfer speed.

Several test campaigns were performed to evaluate the variance of the test, but all of them showed similar results. In Fig. 1 two representative examples are presented. To calculate the transfer speed, the following formula was used:

$$Transfer\ speed = \frac{Total\ data\ transferred}{Longest\ transfer\ duration}$$

This formula allows calculation of the worst transfer speed of one file during the test campaign. For EOS the calculated transfer speed was 990 GB/s with 200 jobs and for dCache it was estimated to be 1390 GB/s with 176 jobs. It should be mentioned that all the tests were performed on a working infrastructure, so some minor interference may

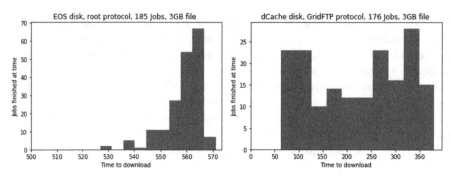

Fig. 1. Data download times in running jobs

be caused by other activities. On the other hand, the demonstrated plots represent real transfers performed under normal conditions.

The measurement of the download speed provides information to estimate how much data can be simultaneously downloaded to the computing element. But it does not give evidence of the most important reasons for the transfer speed limitation. It may be network performance issues, SE architectural performance limitation, bad performance of a hardware, wrong configuration, or heavy use of the storage resources by other jobs. Nevertheless, this test is a good starting point for understanding which components should be optimized to increase the overall performance. It also provides useful numbers for the users who need to download data during their job's execution.

5.2 Performance Test of Computing Resources

The data downloaded to a worker node is processed by the user application. The rate of the data processing impacts the rate of data download. On the other hand, the rate of processing depends on the performance of the CPU. Measurements of computing resources performance are usually based on some synthetic tests chosen to represent most closely typical computational tasks. It is impossible to have a single fully representative benchmark test. The results depend a lot on the properties of the test load and the performance measurements correspond only approximately to the performance with real jobs.

For our purposes, the best approach is to use one of the existing benchmarks. HEP-SPEC2006 is a standard benchmark in computing for high energy physics [15]. But it is mostly used by local administrators of computing resources since it is based on non-free benchmark SPEC2006. The test is rather heavy and takes several hours to run. For our use case, the DIRAC Benchmark 2012 (DB12) turned out to be a good alternative to HEP-SPEC2006 [16]. This benchmark was originally created for prediction of the duration of LHCb Monte-Carlo tasks. It is fast (takes around 60 s), and it runs every time the DIRAC pilot job agent starts execution.

The fact that DB12 benchmark is executed in each pilot job agent allows us to use the already running pilots for the estimation of the resource's performance. We used the test approach described in the previous section: we submit many jobs on one particular computing resource and start all benchmarks simultaneously. This is done to be sure that

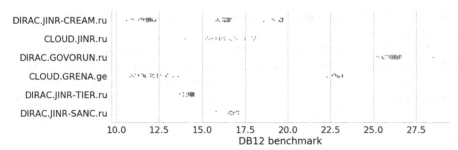

Fig. 2. Results of DB12 benchmark on different resources.

the benchmark on a particular CPU core is performed when other CPU cores are also busy. The benchmark tests in regular pilot jobs cannot guarantee that.

The results are presented in Fig. 2. The tests were performed on several resources. The biggest ones are: Tier1 grid site (DIRAC.TIER.ru), Tier2 grid site (DIRAC.JINR-CREAM.ru), the Govorun Supercomputer (DIRAC.GOVORUN.ru), and the JINR Cloud sites (CLOUD.JINR.ru and others). Other resources are also presented on the plot just for the reference. Every dot is a particular test job result. Dots related to a particular resource are placed randomly within some margin in order to demonstrate the amount of different results with the same value and avoid overlapping. The more to the right the point is - the more performant the resource that it represents.

If one resource has twice better results in DB12 benchmark than another one, it means that the computing task should run on the first resource twice faster. Or, in other words, the first resource should provide twice more results per time period than the second one. The benchmark values represent the performance not of the whole CPU but only one core of it.

The results of this test should not be used to estimate the "power" of a computing resource as a whole because the benchmarks examine only one core of a processor and tell nothing about the number of cores on the resource. To get this information one should take the total amount of job slots provided by the resource and multiply it by the average benchmark result.

To validate the results of DB12 benchmark in our infrastructure we decided to analyze the information about user jobs running under the control of the DIRAC WMS. Every job has information about the performance of the core it is running on. This information

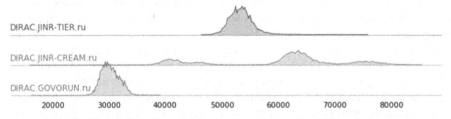

Fig. 3. Distribution of users' jobs duration in seconds on different resources.

is stored in the DIRAC database. We retrieved this information for jobs with similar user payload properties.

In Fig. 3 the duration of the user jobs is presented. The data here corresponds well to the results of dedicated tests presented in Fig. 2. In this test, only Tier1, Tier2 sites and the Govorun Supercomputer were used. Discrepancies here may be due to several reasons: input data transfer delays; output data transfer delays; random variations in the user application execution time (Monte-Carlo simulation).

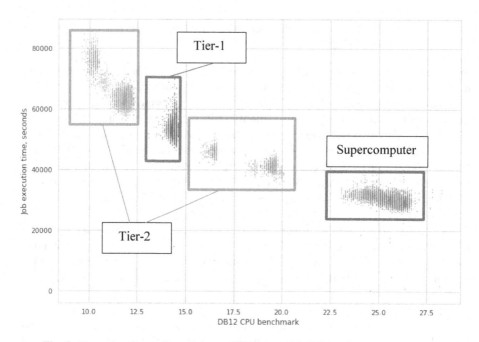

Fig. 4. Execution times of user jobs on CPU cores with different benchmark results.

The same results are presented in Fig. 4 to demonstrate the correlation between the CPU performance benchmark and the real jobs execution time. Each dot in the plot corresponds to a single job. As one can see, all the dots are fitting a hyperbola shape area which shows a strong linear correlation of the job execution speed and the DB12 benchmark of the corresponding CPU. There are distinct clusters of the dots in the plot corresponding to computing clusters with different CPU models having different performance.

The results of the tests show the accuracy of user jobs execution time prediction on a given CPU using the DB12 benchmark provided that the user payload is first characterized by running on some reference processor with a known performance benchmark. The accuracy is estimated to be around 15%. The benchmark-based job execution time estimation can be used by users for monitoring their jobs. In some cases, users can choose resources with higher performance if the job turnaround time is important. The benchmark information can be also used by the DIRAC WMS job scheduler to place jobs with high CPU requirements or higher priority jobs to a more performant computing resource.

6 Conclusion

The Joint Institute for Nuclear Research is a large scientific organization with several big computing and storage subsystems. Most of the time they are used separately by scientific groups and there is no simple way to aggregate them to allow workflows spanning the whole computing center. Different components could be separated by a slower network, different authentication systems and different protocols. This problem becomes more visible when one of the resources is overloaded while others are underused. If a good interoperability between the components could be ensured, it would be possible to easily switch between them in order to balance their loads. There are many tasks that are not bound to use some particular type of resources neither because of technical compatibility nor because of the adopted usage policies.

With the interware technology, it turns out to be easy to integrate computing resources and access them as a single meta-computer. This leads to significant improvements in efficiency of usage of the JINR computing infrastructure. Multiple JINR scientific groups can benefit from the uniform interfaces to various computing facilities. Therefore, integration services based on the software provided by the DIRAC Interware project were set up and evaluated.

The DIRAC services were installed at JINR in order to integrate resources used by big experiments like MPD and BM@N. The initial tests and measurements demonstrated the possibility to use it for data reconstruction and Monte-Carlo generation on all the JINR resources: Grid cluster, Computing Cloud and Govorun supercomputer.

While integrating heterogeneous computing resources, it is still important to keep track of the properties and performance metrics of each component in the integrated system. Therefore, a set of tests to evaluate the performance of different clusters were introduced and validated in a production environment with real users' tasks. The resulting metrics and benchmarks led to a better understanding of the performance of the integrated system as a whole and provided the necessary information for users and for the task meta-scheduling mechanism to optimize the task placement in different computing clusters depending on the task resources requirements and priorities.

References

1. Bozzi, C., Roiser, S.: The LHCb software and computing upgrade for Run 3: opportunities and challenges. J. Phys.: Conf. Ser. **898**, 112002 (2017). https://doi.org/10.1088/1742-6596/898/10/112002
2. SKA Telescope. https://www.skatelescope.org/software-and-computing/. Accessed 19 Aug 2019
3. Kalinichenko, L., et al.: Data access challenges for data intensive research in Russia. Inform. App. **10**(1), 2–22 (2016). https://doi.org/10.14357/19922264160101
4. Korenkov, V., Pelevanyuk, I., Tsaregorodtsev, A.: DIRAC system as a mediator between hybrid resources and data intensive domains. In: Selected Papers of the XXI International Conference on Data Analytics and Management in Data Intensive Domains (DAMDID/RCDL 2019), vol. 2523, pp. 73–84, Kazan, Russia (2019)

5. France Grilles. http://www.france-grilles.fr. Accessed 01 Feb 2020
6. Britton, D., et al.: GridPP: the UK grid for particle physics. Philos. Trans. R. Soc. A **367**, 2447–2457 (2009)
7. European Open Science Cloud. https://www.eosc-portal.eu. Accessed 19 Aug 2019
8. Tsaregorodtsev, A.: DIRAC distributed computing services. J. Phys: Conf. Ser. **513**(3), 032096 (2014). https://doi.org/10.1088/1742-6596/513/3/032096
9. Gergel, V., Korenkov, V., Pelevanyuk, I., Sapunov, M., Tsaregorodtsev, A., Zrelov, P.: Hybrid distributed computing service based on the dirac interware. Commun. Comput. Inf. Sci. **706**, 105–118 (2017). https://doi.org/10.1007/978-3-319-57135-5_8
10. Baginyan, A., et al.: The CMS Tier1 at JINR: five years of operations. In: Proceedings of VIII International Conference on Distributed Computing and Grid-technologies in Science and Education, vol. 2267, pp. 1–10 (2018)
11. Baranov, A., et al.: New features of the JINR cloud. In: Proceedings of VIII International Conference on Distributed Computing and Grid-technologies in Science and Education, vol. 2267, pp. 257–261 (2018)
12. Adam, Gh., et al.: IT-ecosystem of the HybriLIT heterogeneous platform for high-performance computing and training of IT-specialists. In: Proceedings of VIII International Conference on Distributed Computing and Grid-technologies in Science and Education, vol. 2267, pp. 638–644 (2018)
13. Peters, A.J., et al.: EOS as the present and future solution for data storage at CERN. J. Phys.: Conf. Ser. **664**, 042042 (2015). https://doi.org/10.1088/1742-6596/664/4/042042
14. dCache, the Overview. https://www.dcache.org/manuals/dcache-whitepaper-light.pdf. Accessed 19 Aug 2019
15. Michelotto, M., et al.: A comparison of HEP code with SPEC benchmarks on multi-core worker nodes. J. Phys.: Conf. Ser. **219**, 052009 (2010). https://doi.org/10.1088/1742-6596/219/5/052009
16. Charpentier, P.: Benchmarking worker nodes using LHCb productions and comparing with HEPSpec06. J. Phys.: Conf. Ser. **898**, 082011 (2017). https://doi.org/10.1088/1742-6596/898/8/082011 (IOP Conf. Series)

On Information Search Measures and Metrics Within Integration of Information Systems on Inorganic Substances Properties

Victor A. Dudarev[1]([✉]) [iD], Nadezhda N. Kiselyova[2] [iD], and Igor O. Temkin[3] [iD]

[1] National Research University Higher School of Economics, Moscow 109028, Russia
vdudarev@hse.ru

[2] A.A. Baikov Institute of Metallurgy and Materials Science of RAS (IMET RAS),
Moscow 119334, Russia

[3] National University of Science and Technology MISIS
(Moscow Institute of Steel and Alloys), Moscow 119049, Russia

Abstract. One of the main tasks in the integration of information systems is to provide relevant retrieval of information consolidated from heterogeneous sources. In the field of inorganic chemistry and materials science, set-theoretic methods of searching for relevant information are known. They ensure the construction of a sufficiently high-quality response to user requests. However, the problem of quantifying evaluation of information search relevance in this subject area remains open. This paper proposes an approach to quantifying evaluation of the relevance of information retrieval in integrated systems on inorganic substances and materials properties by introducing the relevance graph built on chemical objects. A "chemical similarity" metric and measure are proposed and their properties are discussed.

Keywords: Relevance evaluation · Chemical similarity metrics · Database integration · Inorganic substances

1 Introduction

The development and application of integrated information systems, on substances and materials properties, that consolidate information from heterogeneous information sources, is a worldwide common trend. These systems ensure that specialists are capable of quickly finding the required information. When developing such systems, the fundamental principle is the data representation method which describes corresponding chemical objects and their properties. Furthermore, chemical objects data representation method, in its turn, determines the class of methods for ensuring the search for relevant information and their functionality. The purpose of this paper is to present a new approach for quantifying the evaluation of the relevance of information retrieval, for integrated information systems (IS) on inorganic substances and materials properties (ISMP) based on information structures describing the qualitative and/or quantitative

© Springer Nature Switzerland AG 2020
A. Elizarov et al. (Eds.): DAMDID/RCDL 2019, CCIS 1223, pp. 47–58, 2020.
https://doi.org/10.1007/978-3-030-51913-1_4

substance composition. The main idea, to use weighted graph for chemical objects representation, was taken from the conference paper [1] and here is significantly extended by introducing means to consider chemical systems power. The resulted graph is more balanced regarding subgraphs within particular chemical system objects. Other thoughts, regarding resulted "chemical similarity" measures and metrics and their usage within the integrated IS PISM, were considered.

2 The Current State of the Problem

2.1 Heterogeneous Information Systems

Information technology development and the emergence of powerful hardware and software tools for storing and processing information, stimulates works on information systems development in the field of inorganic materials science. As a result, a large number of highly specialized information systems have been developed that are focused on solving problems with due regard for specificity, conditioned by a specific subject domain and research areas of a specific organization developing IS. An example is a number of information systems, based on databases, developed and maintained by IMET RAS. The IMET RAS information systems core consists of a number of databases which store data about a variety of substances properties:

- «Diagram» – database (DB) on the phase diagrams of semiconductor systems;
- «Crystal» – DB on the properties of acoustooptical, electro-optical and nonlinear-optical substances;
- «Phases» – DB on the general properties of ternary and quaternary compounds;
- «Bandgap» – DB on the band gap of inorganic substances [2];
- «Elements» – DB on the properties of chemical elements.

These databases are heterogeneous not only by data structures, but also by software and hardware tools ensuring their operation [3]. It should be noted that above mentioned DBs contain extensive information, but in a fairly narrow area. The situation, when none of the developed information systems contains a complete set of data, on properties of an object (substance or material), and the specialist needs to use several information resources at once, to search for the necessary information, is typical, not only for inorganic materials science, but also for other subject domains.

Obviously, to ensure a high-quality information service for materials scientists, information systems integration is necessary in this subject domain. In Russia, the first successful attempts in this direction were undertaken, at the beginning of the century, at the IMET RAS, for the integration of information systems, mostly used by Russian users [4]. The integration allowed a consolidation of information resources for end users and a significant reduction of the time spent by specialists to find the necessary information. The applied consolidation approach was based on the Enterprise Application Integration (EAI) method and showed its efficiency and good scalability when connecting resources developed in different organizations to the integrated information system. For example, «Thermal Constants of Substances» (TCS), reference book, developed, by the Joint Institute for High Temperatures of Russian Academy of Sciences (JIHT RAS), together with the Moscow State University (MSU)) and «AtomWork» (information system on inorganic substances properties, developed by the National Institute for Materials Science (NIMS), Japan) are among the most successfully integrated systems [5].

One of the main difficulties in the heterogeneous information system (IS) integration is the diversity of the chemical objects described in them. So, for example, «Diagram» IS contains information at the level of the chemical system, i.e. a set of chemical elements that form a certain phase diagram of a semiconductor system. Other IS on inorganic substances and materials properties (ISMP) describe the properties at a specific quantitative composition level (with a specific ratio of elements in the chemical system), taking into account crystal modifications of substances, i.e. the quantitative composition of the substance and its crystal lattice are described at this level. Such chemical objects descriptions incompatibility, in different IS ISMP, dictates the need to use a different description of chemical objects in an integrated IS ISMP, to distinguish between several types of chemical objects: chemical systems, substances and their crystal modifications.

2.2 Chemical Objects Hierarchy

To describe the basic chemical objects of the considered problem domain, the set theory is used, taking into account that each subsequent level in the problem domain hierarchy complements the description of the chemical object. The notation is the following: S is the set of chemical systems; C – set of chemical substances, i.e. chemical compounds, solid solutions, heterogeneous mixtures, etc.; M – set of crystal modifications. Then the chemical system is denoted as s (where $s \in S$), the chemical substance is denoted by c (where $c \in C$), and the crystal modifications is m (where $m \in M$).

Having designated second level objects by the «substance» term, we get three-level chemical objects hierarchy: chemical system, chemical substance and chemical modification [6]. As far as information stored in DBs on inorganic substances properties can be considered at chemical system level, for simplicity we'll use this level, from the top of the objects hierarchy. So, the chemical objects hierarchy and relationships between chemical objects can be described by means of chemical objects hierarchy in a tree form (Fig. 1).

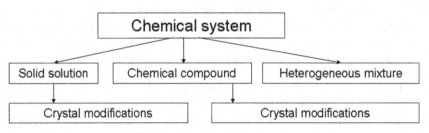

Fig. 1. Chemical objects hierarchy.

Any chemical system s can be represented as a set of chemical elements e_i: $s = \{e_1, e_2, ..., e_n\}$. Any chemical substance, c, is defined not only by the set of atoms (chemical elements), but also by their quantitative incorporation into the composition of the compound, solution or mixture. Therefore, any substance, c, can be represented by a tuple, (s, f), where $s \in S$, and f is a mapping of the set of atoms (chemical elements) that make up the substance, in the set of $\boldsymbol{R}^* \times \boldsymbol{R}^*$ pairs that define the minimum and maximum incorporation of a given chemical element in a compound, solution or mixture, c.

That is, $f: e_i \rightarrow (\boldsymbol{R}^*_{min}, \boldsymbol{R}^*_{max})$, where $\boldsymbol{R}^* = \boldsymbol{R}^+ \cup \{x\}$. $\boldsymbol{R}^+$ – is the set of non-negative real numbers, and $\boldsymbol{R}^*$ is the set of $\boldsymbol{R}^+$ extended by the element, x. The element, x, is used to denote an unknown number, since, in the notation of mixtures, where the incorporation of components may vary, it is customary to use x to denote an unknown, for example, $Fe_{1-x}Se_x$. $\boldsymbol{R}^*_{min}, \boldsymbol{R}^*_{max}$ are, respectively, the minimum and maximum concentration of the chemical element, e_i, in the substance, c.

In the case when the concentration of a particular chemical element, e_i, in the substance, c, is fixed, then $\boldsymbol{R}^*_{min} = \boldsymbol{R}^*_{max}$. Chemical modification, m, can be represented by a tuple (s, f, mod), where $s \in S, f: e_i \rightarrow (\boldsymbol{R}^*_{min}, \boldsymbol{R}^*_{max})$, and mod is the string notation for the crystal modification of a substance – common for integrated IS ISMP (one of the singony enumeration values: {*Triclinic, Monoclinic, Orthorhombic, Tetragonal, Trigonal, Hexagonal, Cubic*}).

2.3 Metabase Structure

Quite reasonably, when designing integrated IS ISMP, it's required to provide search facilities for relevant information, contained in other IS ISMP of a distributed system. Therefore, it's required to develop some active data store that should "know" what information is contained in every integrated IS ISMP. Considering chemical objects hierarchy, some database should exist that describes information contained in integrated resources in terms of chemical systems, substances and crystal modifications. Here we come to the metabase concept – a special database that contains metadata that describe integrated IS ISMP contents in terms of chemical objects hierarchy, as well as some additional information on users and their permissions, together with information required to integrate distributed IS ISMP (Fig. 2).

The metabase defines integrated IS capabilities. Its structure should be flexible enough to represent metadata on integrated ISs ISMP contents and, at the same time, the metabase structure should be simple and versatile to describe the arbitrary data source, on inorganic substances properties, without exhaustive additional payload, currently offered by numerous materials ontologies. Taking into consideration the fact that chemical objects and their corresponding properties description is given at different detail level in different ISs ISMP, it's important to develop a metabase structure that would be suitable for the description of information, residing in different ISs ISMP. For example, some integrated DBs contain information on particular crystal modifications properties, while others contain properties description at chemical system level. Thus, integrated ISs ISMP deal with different chemical objects, situated at different chemical objects hierarchy levels. For simplicity, in the current paper, we consider only a part of metabase structure, that is devoted to chemical systems and their properties (Fig. 2). The amount of this metainformation should be enough to perform a search for relevant information on systems and corresponding properties.

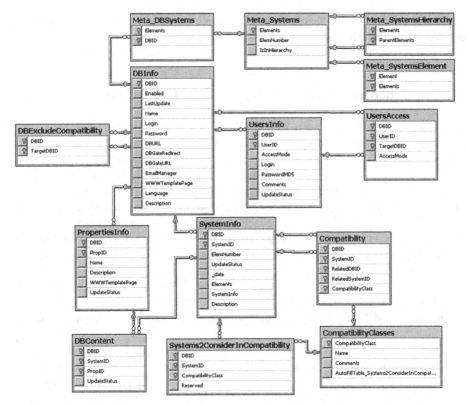

Fig. 2. A Part of the metabase logical structure on the chemical systems level only.

All tables (Fig. 2) can be logically separated into several groups according to their purpose:

- DBInfo – root table, that contains information on integrated database systems;
- DBExcludeCompatibility – table that stores exception list of ISs for relevant information search;
- UsersInfo, UsersAccess – tables that contain information on integrated system users and their access rights to integrated IS ISMP;
- SystemInfo, PropertiesInfo, DBContent – tables that describe contents of integrated IS ISMP;
- CompatibilityClasses, Compatibility, Systems2ConsiderInCompatibility – tables that contain information on accessible relevance classes and their contents (currently 3 relevance classes are used [5]).
- Meta_Systems, Meta_DBSystems, Meta_SystemsHierarchy, Meta_SystemsElement – tables to describe all chemical systems, contained within integrated IS ISMP, with respect to their relation to each other and the chemical elements they consist of.
- Versions – service table (not shown on diagram). It is used for database schema update and versioning.

Taking into account the chemical objects hierarchy description, a special method was developed to search for relevant information in the context of an integrated information system, based on a set-theoretic approach [6].

3   A Set-Theoretic Approach to Relevance Evaluation

Relevance, itself, and its notion to information search, is a philosophic term, covered in numerous publications. A comprehensive review of relevance, itself, is given by Tefko Saracevic [7]. We consider information search relevance, in application to integrated IS ISMP, is close to "chemical similarity" [8]. So, considering the chemical objects hierarchy description, a special method was developed to search for relevant information in the context of an integrated IS ISMP, based on a set-theoretic approach [6]. The main essence of the set-theoretic approach is in the use of abovementioned metabase structure, that is a special database that contains information on integrable IS ISMP (set D), chemical systems (set S) and their properties (set P). To describe the relationship between the elements of the sets D, S, and P, the ternary relation, W, was defined on the set, U (universum), which is the Cartesian product: $U = D \times S \times P$. The element, (d, s, p), belongs to the relation, W, where $d \in D$, $s \in S$, $p \in P$ is interpreted as follows: "the integrable IS ISMP, d, contains information on the p property of the chemical system, s".

Thus, according to accepted notation, the search for relevant information on a particular chemical system, s, can be reduced to proper definition of an R relation, which is a subset of the $S \times S$ Cartesian product (in other words, $R \square S^2$). Thus, for any pair $(s_1, s_2) \in R$, we can state that the s_2 system is relevant to the s_1 system. For the practical solution of the problems of searching for relevant information in integrable information systems, the following rules are often used to construct R [4]:

1. For any set $s_1 \in S$, $s_2 \in S$, which includes the notation of chemical elements e_{ij}, $s_1 = \{e_{11}, e_{12}, ..., e_{1n}\}$, $s_2 = \{e_{21}, e_{22}, ..., e_{2n}\}$, it's true, that if $s_1 \subseteq s_2$ (that is, all chemical elements from s_1 system are contained in s_2 system), then $(s_1, s_2) \in R$.
2. The relation R is symmetric. In other words, for any $s_1 \in S$, $s_2 \in S$ it is true that, if $(s_1, s_2) \in R$, then $(s_2, s_1) \in R$.

It should be noted that the abovementioned automatic variant of R relation generation is just one of the simplest and most obvious variants of such rules, and, in fact, more complex mechanisms can be used to get R relation. Other alternatives are used to build the R relation, called *relevance classes*. For example, browsing information on a particular property of a compound in one of integrated IS ISMP (in fact, it is information defined by (d_1, s_1, p_1) triplet), we consider (d_2, s_2, p_2) triplet to be relevant information. (d_2, s_2, p_2) triplet characterizes information on some other property of a chemical system from another integrated IS ISMP. This enables us to define relevant information more precisely, e.g. if we consider the R relations in the form: $R \square (d_1, s_1, p_1) \times (d_2, s_2, p_2)$, where $d_1, d_2 \in D$, $s_1, s_2 \in S$, $p_1, p_2 \in P$. Actually, it's possible to define a set of several R relations $(R_1, R_2, ..., R_n)$ by applying different rules, to enable users to perform a search for relevant information, based on a wide variety of R interpretations. However complex interpretations of R ($R \square$ (d_1, s_1, p_1) $\times$ (d_2, s_2, p_2)) are not currently being used in the IMET RAS, since the metabase structure would be more complex to store such relations, however, its reasonability is unclear. In the IMET RAS, simple relevancy relations of $R \square S^2$ are used. More rules to form *relevance classes* are given in [5].

Improvement of the search relevance can also be achieved by using the c_i level, i.e. taking into account the quantitative composition of a substance, or crystal modifications of a specific substance, m_i, instead of chemical system designations, s_i, in cases when a user requests relevant information, being at the level of inorganic substances, or their modifications, in the system-substance-modification hierarchy concepts [6].

When searching at the substance level, the quantitative compound composition is taken into account. The pair (a_{imin}, a_{imax}) denotes the quantitative inclusion of chemical element, $e_i \in s$, into the composition, a_{imin}, $a_{imax} \in R^+$, $a_{imin} \leq a_{imax}$. If $a_{imin} = a_{imax}$, then the substance has a constant composition by the element, $e_i \in s$. For each element of the chemical system, $e_i \in s$, the user during the search could specify a pair, (r_{imin}, r_{imax}), where r_{imin}, $r_{imax} \in \boldsymbol{R^+}$, denoting the allowable interval of the i-th element in the substance (R^+ is the set of non-negative real numbers). Then, all substances, belonging to the same chemical system, are considered relevant, if, for each pair (r_{imin}, r_{imax}), the following is correct: $a_{imin} \in [r_{imin}, r_{imax}]$ or $a_{imax} \in [r_{imin}, r_{imax}]$. In other words, if the logical disjunction $[r_{imin} \leq a_{imin}$ & $a_{imin} \leq r_{imax}] + [r_{imin} \leq a_{imax}$ & $a_{imax} \leq r_{imax}] =$ *true* for all $e_i \in s$, then the data on the substance are considered relevant.

When searching for relevant information, taking into account the crystal modifications of m_i, crystal systems, information on crystal structures is shown in different ways. For example, for lithium niobate ($LiNbO_3$) a hexagonal or trigonal crystallographic system is indicated in different information sources of the IS ISMP [9], which, in fact, corresponds to the same crystal modification.

However, it should be noted that, despite the fact that the described approach, in general, provides an acceptable level of search relevance for inorganic compounds, it suffers from the inability to obtain a quantitative assessment of the search relevance and, as a consequence, the fundamental inability of changing the search results order, by adjusting some parameters or corresponding metrics. Note that such an adjustment is useful in some cases, in particular when preparing training data sets for machine learning tasks in computer-aided construction of inorganic compounds [10].

4 Graph Approach to Relevance Assessment

To search for relevant information and obtain a quantitative measure of relevance assessment, within an integrated information system, based on the properties of inorganic substances and materials, we propose to use a graph model, based on the weighted graph $G = (V, E)$, built on chemical objects, described as part of an integrated information system.

Let's define a set of vertices, V, for graph, G. In accordance with the accepted three-level description of chemical objects in an integrated information system, the set of vertices consists of three disjoint subsets, $V = \{S, C, M\}$, where S is the set of chemical systems s_i (qualitative compound composition), C is the set of chemical compounds c_i (the quantitative compound composition or the substance formula), M is the set of crystal modifications m_i of specific substances.

Define a set of edges, E, for graph, G, as the union of non-intersecting subsets $E = Es \cup Ec \cup Em \cup Esc \cup Ecm$, where Es – edges that are incidental only to the set of vertices S; Ec – edges that are incidental only to the set of substances C; Em – the edges that are incidental only to modifications set M. The vertices connectivity for the classes of S, C, M is achieved by two sets of edges: Esc edges to connect vertices from S and C sets; and Ecm edges to connect vertices from C and M. Please note that the edges connecting vertices from S and M sets, are absent.

To define the elements of the E subsets, we need to introduce a couple of trivial functions: $Fs(c)$ and $Fc(m)$. The $Fs(c)$ function returns the chemical system for a given compound, c, i.e. it allows us to get qualitative composition from quantitative composition. The $Fc(m)$ function returns quantitative composition of a particular crystal modification of the substance, i.e. it allows us to get quantitative composition from a particular crystal structure of the compound. Then, given that the chemical system is a set of chemical elements, $s = \{e_1, e_2, ..., e_n\}$ we get the following set of edges:

$$Es = \{(s_i, s_j)\}, \text{ where } s_i = \{e_{i1}, e_{i2}, \ldots, e_{in}\}, s_j = \{e_{j1}, e_{j2}, \ldots, e_{jm}\}, |s_i| = n, |s_j| = m, m - n = 1, s_i \Box s_j; \tag{1}$$

$$Ec = \{(c_i, c_j)\}, \text{ where } Fs(c_i) = Fs(c_j); \tag{2}$$

$$Em = \{(m_i, m_j)\}, \text{ where } Fc(m_i) = Fc(m_j); \tag{3}$$

$$Esc = \{(s_i, c_j)\}, \text{ where } Fs(c_j) = s_i; \tag{4}$$

$$Ecm = \{(c_i, m_j)\}, \text{ where } Fc(m_j) = c_i. \tag{5}$$

When searching for relevant information on a chemical object, it is necessary that a path should exist in the graph between the initial object and a relevant one, and it is easy to calculate the measure of relevance by adding the weights of the edges on the corresponding path. Thus, we come to the necessity of introducing a real-valued function, W, defined on the graph edges set:

$$W(Es) = W\big((s_i, s_j)\big) = 10^N, \text{ where } N = \max\big(|s_i|, |s_j|\big); \tag{2.1}$$

$$W(Ec) = W\big((c_i, c_j)\big) = \min\Big(\sum_{k=0}^{n} 10^k * |q_{ik} - q_{jk}|\Big); \tag{2.2}$$

where $n = |Fs(c_i)| = |Fs(c_j)|$, q_{ik} and q_{jk} – quantitative occurrence of k-th element at c_i and c_j compositions, i.e. $Q\colon e_k \to R^+$ (respectively $Q(el_{ik}) = q_{ik}$, $Q(e_{jk}) = q_{jk}$), and the order of elements for formula calculation should be selected in the way to ensure the minimum value of the $W(Ec)$ objective function.

$$W(Em) = 0.1; \tag{2.3}$$

$$W(Esc) = W\big((s_i, c_j)\big) = 10^N, \text{ где } N = |Fs(c_j)| = |s_i|; \tag{2.4}$$

$$W(Ecm) = 0.5. \tag{2.5}$$

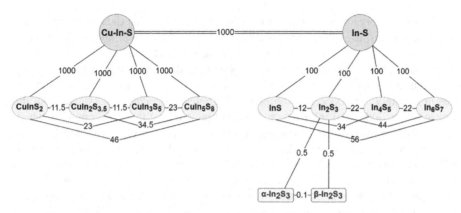

Fig. 3. Fragment of the relevance graph for Cu-In-S and In-S chemical systems.

As an example, consider a relevance graph fragment for Cu-In-S and In-S chemical systems (Fig. 3). In this example, we emphasize its properties and justify the role of edge weights for quantitative assessment of the chemical objects' relevance.

Based on the definition of E edges set, it can be seen that the relevance graph is partitioned into subgraphs, based on the vertices from S (chemical systems) set. Moreover, there is no path in the graph between substances from different chemical systems, bypassing chemical systems vertices. The systems vertices, themselves, are connected by an edge only if the elements set of the first system is an own subset of the second system and their powers (i.e. chemical elements count that built up a system) differ by one.

Consider a subgraph constructed on the basis of the In-S chemical system vertex and consisting of substances and their corresponding modifications related to this system. It should be noted that the subgraph, composed on vertices of a C set (compounds, i.e. qualitative formula) is complete, since all the vertices (InS, In_2S_3, In_4S_5, In_6S_7) are connected to each other and form a clique. Note, however, that the weights of the edges, connecting the substances' vertices, are different. Edge weight is a number characterizing the degree of closeness of corresponding quantitative compositions: the smaller the difference, the lower the weight («cost») of transition along the edge, and the corresponding substance is considered more relevant than the other with greater path weight value.

Similarly, modifications subgraph, constructed on the basis of the vertex designating a particular compound, is complete, and the all edges weights are equal to 0.1. In Fig. 3 such edges are connected to each other, e.g. these are α-In_2S_3 and β-In_2S_3 vertices. Note, that the transition in the graph from modification to the corresponding substance has a cost of 0.5, and the transition from substance to the system – 100 (10^2), which makes more relevant data on other modifications (including crystal structure) than the transition to the substances level to choose another qualitative composition.

As it can be easily seen, the W function (2.1–2.5), defined on the set formed by the elements of S, C and M sets, satisfies the identity, symmetry and triangle inequality axioms. Thus, the W function is a *metric* in the chemical objects space.

The IS on inorganic substances properties peculiarity is that data, on various substances properties, are often stored at different levels of chemical objects description. For example, in the «Crystal» IS, data on substances refractive indices are stored at the level of crystal modifications (M), at the same time data on melting temperature are defined at the quantitative substances composition (C) level. Moreover, «Crystal» IS does not contain the information on phase diagrams of the chemical systems, but the corresponding additional information can be obtained from the «Diagram» IS, which describes this property at the qualitative composition level of chemical objects, i.e. at the chemical systems (S) level. Thus, in order to provide all data on particular chemical object properties, it is necessary to combine the properties set at higher levels of description. Therefore, the edges connecting the vertices of different classes (*Esc* and *Ecm*) should be converted into pairs of arcs (arrows) with different weights so that the arcs weights from modifications to substances and from substances to systems are equal to zero:

$$W'(\{(c_i, m_j)\}) = 0.5, \text{ where } Fc(m_j) = c_i; \tag{3.1}$$

$$W'(\{(m_j, c_i)\}) = 0, \text{ where } Fc(m_j) = c_i; \tag{3.2}$$

$$W'(\{(s_i, c_j)\}) = 10^N, \text{ where } Fs(c_j) = s_i, N = |Fs(c_j)|; \tag{3.3}$$

$$W'(\{(c_j, s_i)\}) = 0, \text{ where } Fs(c_j) = s_i. \tag{3.4}$$

Obviously, the W' function (3.1–3.4) is a *measure* in the chemical objects space, which makes it possible to discover the full chemical object properties set – to achieve this it is enough to obtain the properties of all objects that can be reached with zero cost path. Note that W' is not a metric due to the violation of symmetry and triangle inequality rules. The oriented graph of chemical properties relevance (Fig. 4) contains arcs with zero weight, denoted by a dotted curve, added according to (3.2) and (3.4) rules.

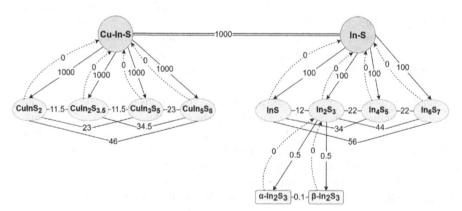

Fig. 4. Fragment of the oriented graph of relevance properties for Cu-In-S and In-S chemical systems.

5   Discussion and Further Model Development

The proposed graph model is an attempt to reflect the similarity degree of various chemical objects even described at different representation level (system, compound, modification). In this sense, the path cost is a difference measure between the corresponding chemical objects, which are the graph vertices. The more similar the objects, the «closer» they are, meaning the path cost in the graph is less. It is planned to use the measure and metric proposed in the work when ranking the relevant information displayed to user in integrated IS on inorganic substances properties, in particular in the single access point http://www.imet-db.ru, given that the relevance is inversely proportional to the graph path cost.

It is worth noting that, in a broad sense, according to the definitions, given in the paper, the overall relevance graph is disconnected due to the absence of a path between the vertices of chemical systems, that have no common chemical elements (i.e. $s_1 \in S$, $s_2 \in S$ such that $s_1 \cap s_2 = \emptyset$). For example, in the current model, there is no connectivity between In-S and Ga-As chemical systems, although In and Ga are similar in many ways, as far as In and Ga are elements from the same subgroup of the periodic system. In this case, it makes sense to introduce, in future, rules for additional edges formation between similar substances and systems (in which an element from the same periodic system subgroup changes), although such an edge should have an appropriate (sufficiently high) weight comparing with analogues with common chemical elements.

In tr o du cti on of measure W' allows us to select several graph edges in order to collect all the information on the requested chemical object from different IS on inorganic substances properties, which is especially useful when building data sets for machine learning or when searching for all the properties of the chemical object required.

As possible ways of further graph model development, it's planned further edges weight adjustment according to (2.1)–(2.5) rules considering not only the power of chemical system set but other factors derived from the Periodic Law. This should contribute to the more accurate graph balancing with respect to the multicomponent inorganic systems.

6   Conclusion

In the paper by means of the graph model, the relevant information search concept was extended regarding to integrated IS ISMP. The new model allows us to obtain quantitative information retrieval relevance assessment, based on the path calculation in the weighted graph, which allows ranking of chemical information found in the consolidated data sources. By means of edges weights in the chemical objects space, a metric is introduced that enables us to quantify the similarity degree of chemical objects, belonging to various classes (system, compound, crystal modification). The proposed approach is applicable not only to improve information retrieval for end users – material chemists, but also for application in computer-aided design of inorganic compounds at the stage of training samples formation, based on the quantitative relevance assessment.

Acknowledgement. This work was partially supported by the Russian Foundation for Basic Research (project no. 18-07-00080) and the State task № 075-00746-19-00.

References

1. Dudarev, V.A., Kiselyova, N.N., Temkin, I.O.: Relevance Evaluation of Information Retrieval in the Integration of Information Systems on Inorganic Substances Properties. Selected Papers of the XXI International Conference on Data Analytics and Management in Data Intensive Domains (DAMDID/RCDL 2019). Kazan, Russia, 15–18 October 2019. CEUR Workshop Proceedings, vol. 2523, pp. 348–357. http://ceur-ws.org/Vol-2523/paper34.pdf
2. Kiselyova, N.N., Dudarev, V.A., Korzhuyev, M.A.: Database on the bandgap of inorganic substances and materials. Inorganic Mater.: Appl. Res. 7(1), 34–39 (2016)
3. Kiseleva, N.N., et al.: Database system on materials for electronics on the Internet. Inorganic Mater. 40(3), 380–384 (2004)
4. Kornyshko, V.F., Dudarev, V.A.: Software development for distributed electronics materials. In: Proceedings of the Third International Conference "Information Research, Applications and Education - i.Tech ", Sofia, FOI-Commerce, pp. 27–33 (2005)
5. Dudarev, V.A., Kiselyova, N.N., Xu, Y., Yamazaki, M.: Virtual integration of the Russian and Japanese databases on properties of inorganic substances and materials. In: MITS 2009. In Proceedings of Symposium on Materials Database, National Institute for Materials Science (NIMS), Materials Database Station (MDBS), pp. 37–48 (2009)
6. Dudarev, V.A.: Integration of information systems in the field of inorganic chemistry and materials science. ISBN 978-5-396-00745-1, M.: KRASAND, 320 p. (2016)
7. Saracevic, T.: Relevance: a review of the literature and a framework for thinking on the notion in information science. Part II: nature and manifestations of relevance. J. Am. Soc. Inf. Sci. Technol. 58(3), 1915–1933 (2007)
8. Johnson, A.M., Maggiora, G.M.: Concepts and Applications of Molecular Similarity, p. 393. Wiley, New York (1990). ISBN 978-0-471-62175-1
9. Serain, D.: Middleware and Enterprise Application Integration, p. 288. Springer, London (2002). ISBN 978-1-85233-570-0
10. Sen'ko, O.V., Kiselyova, N.N, Dudarev, V.A., Dokukin, A.A., Ryazanov, V.V.: Various machine learning methods efficiency comparison in application to inorganic compounds design. In: Selected Papers of the Data Analytics and Management in Data Intensive Domains. Proceedings of the XX International Conference – DAMDID/RCDL 2018, 9–12 October 2018, Moscow, vol. 2277, pp. 152–158 (2018)

Models, Ontologies and Applications

Models Are Functioning in Scenarios

Bernhard Thalheim$^{(\boxtimes)}$

Department of Computer Science, Christian-Albrechts University at Kiel,
24098 Kiel, Germany
thalheim@is.informatik.uni-kiel.de
http://bernhard-thalheim.de/

Abstract. Models are one of the universal instruments of humans. They
are instruments that are used on purpose in scenarios. They are usable
and useful within these scenarios. The scenario determines the function
and thus the purpose of a model. This determination governs the acceptance in dependence of the utility that a model provides.

We discuss in this paper the dependence of model kinds on four scenarios: communication scenario oriented on social aspects, comprehension
scenarios targeting on representation aspects, search scenarios oriented
on model usage and model enhancement, and analysis scenarios targeting
on sense-making aspects.

1 Introduction

Models are widely used in life, technology and sciences. We claim that models are one of the first instruments human use. For instance, newborn children
develop their own concept of 'mother' or 'father' based on observations. Models
are subconscious or preconscious and have no proper language representation.
Language is an instrument that is developed later and then widely used. Models also become then language-based. At the early stage, models are oriented
on interaction and social aspects. Later models are oriented on representation,
foundation and sense-making, or realisation and context aspects.

The later usage of models is already driven by these three additional aspects.
Models will be more or less systematically developed. They become conscious
constructions. The development of models in science and technology is still a
mastership of an artisan and not yet systematically guided and managed. Models allow model-based reasoning. The main advantage of model-based reasoning
is based on two properties of models: they are focused on the issue under consideration and are thus far simpler than the application world and they are reliable
instruments since both the problem and the solution to the problem can be
expressed by means of the model due to its dependability.

Models in science and technology must be sufficiently comprehensive for the
representation of the domain under consideration, efficient for the solution computation of problems, accurate at least within the scope, and must function
within an application scenario.

© Springer Nature Switzerland AG 2020
A. Elizarov et al. (Eds.): DAMDID/RCDL 2019, CCIS 1223, pp. 61–81, 2020.
https://doi.org/10.1007/978-3-030-51913-1_5

The Notion of Model

Let us first briefly repeat our approach to the notion of model:

A **model** *is a well-formed, adequate, and dependable instrument that represents origins and that functions in utilisation scenarios* [7,23,25].

Its criteria of well-formedness, adequacy, and dependability must be commonly accepted by its community of practice within some context and correspond to the functions that a model fulfills in utilisation scenarios.

The model should be well-formed according to some well-formedness criterion. As an instrument or more specifically an artifact a model comes with its *background*, e.g. paradigms, assumptions, postulates, language, thought community, etc. The background its often given only in an implicit form. The background is often implicit and hidden.

A well-formed instrument is *adequate* for a collection of origins if it is *analogous* to the origins to be represented according to some analogy criterion, it is more *focused* (e.g. simpler, truncated, more abstract or reduced) than the origins being modelled, and it sufficiently satisfies its *purpose*.

Well-formedness enables an instrument to be *justified* by an empirical corroboration according to its objectives, by rational coherence and conformity explicitly stated through conformity formulas or statements, by falsifiability or validation, and by stability and plasticity within a collection of origins.

The instrument is *sufficient* by its *quality* characterisation for internal quality, external quality and quality in use or through quality characteristics [22] such as correctness, generality, usefulness, comprehensibility, parsimony, robustness, novelty etc. Sufficiency is typically combined with some assurance evaluation (tolerance, modality, confidence, and restrictions).

A well-formed instrument is called *dependable* if it is sufficient and is justified for some of the justification properties and some of the sufficiency characteristics.

Models Function in Scenarios for Which They Are Build

A scenario may be considered to be an application game (in the sense of Wittgenstein's "language game" [31]). It can often be normalised as a graph of activities or practices. The model has to function in such practices. It plays a role. Scenarios often stereotyped and follow conventions, customs, exertions, habits. The scenario determines which instruments can be properly used, which usage pattern or styles can be applied, and which quality characteristics are necessary for the instruments used in those activities.

The typical usage of a model is that of being a deputy for origins that nt directly usable. They are often too complex, too difficult for direct use, too expensive for repeated usage, too fuzzy due to the early development phase, too hidden or closed and thus not directly deplorable, etc. Origins can also be artifacts. They can also be ideas on which should be investigated or build. In computer science, models represented the augmentation of a given reality.

Models are used for engineering a solution to problems that can be solved by such augmentation.

The usage of models is embedded into such practices or more formally scenarios. The scenario or a bundle of scenarios determine the portfolio for an instrument:

- When, how in general, what for, by whom, for whom an instrument and a model as an instrument is used.
- In which function the model is going to be used. what is the resulting purpose (wherefore), what as e the goals. What constitutes the portfolio.
- What is the intended worthiness (mission, determination, ideal and desired identity) of the model.
- In what way, with what requirements must be the model developed.
- What, whereof is deputed/represented, in which scope and focus, at which granularity and precision.
- What adequacy and dependability is required for the deputation by a model.
- Wherewith, on which basis and grounding models can be used.

An instrument such as a model has its usefulness in such scenarios. Due to this usefulness, it becomes usable and has to function according to the needs.

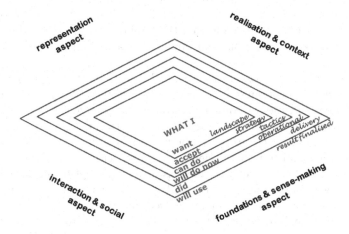

Fig. 1. Aspects of model functioning in scenarios

The four aspects in Fig. 1 are the representation aspect, realisation and context aspect, the foundational and sense-making aspect, and the interaction and social aspects. A function of a model is typically a combination of functions according to these aspects. In general, model development can be characterised by six layers: initialisation by exploring the landscape, development of the strategy, exploration of tactics, operationalisation, model construction, and model finalisation. Each of these layers adds quality characteristics and specific activities to the previous layer in dependence on the aspects considered. Model development must not explicitly follow this layering. The layering allows however to answer questions such as: which model is wanted; which model can become accepted; which models should be constructed; which model realisation is now taking place; which model has been constructed; which model and its variants are delivered.

Model Deployment Scenarios are Multi-facetted

The model notion can be seen as an initialisation for more concrete notions. We observe that model utilisation follows mainly four different kinds of scenarios. The four scenarios do not occur in its pure and undiffused form they are interleaved. We can however distinguish between:

Problem solving scenarios: Problem solving is a well investigated and well organised scenario (see, for instance, [1,9]). It is based on (1) a problem space that allows to specify some problem in an application in an *invariant* form and (2) a solution space that *faithfully* allows to back-propagate the solution to the application. We may distinguish three specific scenarios: perception & utilisation; understanding & sense-making, and making your own.

Engineering scenarios: Models are widely used in engineering. They are also one of the main instruments in software and information systems development, especially for system construction scenario. We may distinguish three specific scenarios depending on the level of sophistication: direct application: managed application, and application according to well-understood technology.

Science scenarios: Sciences have developed a number the distinctive form in which a scenario is organised. Sciences make wide use of mathematical modelling. The methodology of often based on specific moulds that are commonly accepted in the disciplinary community of practice, e.g. [1]. We may distinguish three specific scenarios: comprehension, computation and automatic detection for instance in data science, and intellectual adsorption.

Social scenarios: Social scenarios are less investigated although cognitive linguistics, visualisation approaches, and communication research have contributed a lot. Social models might be used for the development of an understanding of the environment, for agreement on behavioural and cultural pattern, for consensus development, and for social education. We may distinguish three specific scenarios: development of social acceptance, internalisation & emotional organisation, and concordance & judgement.

Figure 2 refines the four aspects of mentioned scenarios introduced in Fig. 1. The notion of model mainly reflects the initialisation or landscape layer. Depending on the needs and demands to model utilisation we may distinguish various layers from initialisation towards delivery. The strategy, tactics, operational, and delivery layers are essentially refinements and extensions of the initialisation. The dependability and especially the sufficiency are based on other criteria while the landscape layer is permanent for all models due to the consideration of the concern, the issue, and the specific adaptation to the community of practice. The

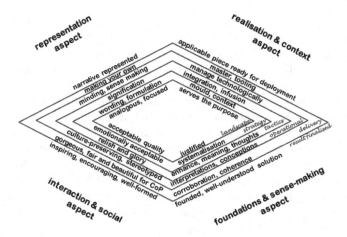

Fig. 2. The aspects of problem solving, engineering, science, social scenarios

strategy layer is governed by the context (e.g. the discipline) and the mould for model utilisation, and the matrix (including methodologies and commonly accepted approaches to modelling). The tactics layer depends on the settlement of the strategy and initialisation layers considers the well-acknowledged experience (e.g. generic approaches), the school of thought or more generally the background, and the framing of the modelling. Which origin(al)s are reflected and which are of less importance is determined in the operational layer that orients on the design and on mastering the modelling process. Finally the model is delivered and form for its application in scenarios that are considered. We thus observe various specific quality characteristics for each of these aspects and layers.

Do We Need a Science of Models and Modelling?

Since everybody is using models and has developed a specific approach to models and modelling within the tasks to be solved, it seems that the answer is "no". From the other side we deeply depend on decisions and understandings that are based on models. We thus might ask a number of questions ourselves. Can models be misleading, wrong, or indoctrinating? Astrophysics uses a Standard Model that has not been essentially changed during the last half century. Shall we revise this model? When? What was really wrong with the previous models? Many sciences use modelling languages in religious manner, e.g. think about UML and other language wars. What is the potential and capacity of a modelling languages? What not? What are their restrictions and hidden assumptions? Why climate models have been deeply changing and gave opposite results compared to the previous ones? Why we should limit our research on impacts of substances to a singleton substance? What is the impact of engineering in this case? What has been wrong with the two models on post-evolution of open cool mines after deployment in Germany which led to the decision that revegetation is far better than water flooding? Why was iron manuring a disaster decision for the Humboldt stream ocean engineering? What will be the impact of the IPCC/NGO/EDF/TWAS proposal for Solar Radiation Management (SRM) for substantial stratosphere obscuration for some centuries on the basis of reflection aerosols (on silver, sulfate, photophoretic etc. basis)? Why reasoning on metaphors as annotations to models may mislead? Are "all models wrong"[1]?

Developing a science of models and modelling would allow us to answer questions like the following one: What is a model in which science under which conditions for whom for which usage at a given time frame? What are necessary and sufficient criteria for an artefact to become a model? What is the difference between models and not-yet-models or pre-models? What is not yet a model? How are models definable in sciences, engineering, culture, ...? Under which conditions we can rely on and believe in models? Logical reasoning: which calculus? Similarity, regularity, fruitfulness, simplicity, what else (Carnap)? Treatment, development, deployment of models: is there something general in common? Models should be useful! What does it mean? Is there any handling of usage, usefulness, and utility? What is the difference between an object, a model, and a pre-model? What might be then wrong with mathematical models? What is the problem in digging results through data mining methods?

The Storyline of This Paper

Models are the first reasoning and comprehension instruments of humans. Later other instruments are developed. The main one is language. Models then often become language-based if they have to be used for collaboration. Others will remain to be conscious, preconscious or subconscious. Based on the clarification of the given notion of model and a clarification of the model-being we explore in this paper what are the constituents of models, how models are composed, and what are conceptions for model constructions. Since models are used in scenarios and should function sufficiently well in these scenarios we start with an exploration of specific nature of models in four scenarios. We are not presenting all details for a theory of models[2].

This paper is a revised, extended and modified version of [26]. Comparing to the previous version, aspects of model functioning in scenarios are discussed in Introduction, model utilization scenarios are described in a more detailed way in Sect. 2, planes supporting model suite layers are discussed in Sect. 3.

2   Case Study on Some Scenarios for Model Utilisation

Models are used in various *utilisation scenarios* such as construction of systems, verification, optimization, explanation, and documentation. *In these scenarios* they *function* as *instruments* and thus satisfy a number of properties [8,27–29].

---

[1] "All models are wrong. ... Obtain a 'correct' one. ... Alert to what is importantly wrong." [3] We claim: *Models might be 'wrong'. But they are **useful**.*

[2] Collections of papers which are used as background for this paper is downloadable via Research Gate. Notions and definitions we used can be fetched there.

Models for Communication

The model is used for exchange of meanings through a common understanding of notations, signs and symbols within an application area. It can also be used in a back-and-forth process in which interested parties with different interests find a way to reconcile or compromise to come up with an agreement.

Communication scenarios are an acts or instances of transmitting by which information is exchanged between individuals through a common system of symbols, signs, or behavior. It is a technique for expressing ideas effectively Typical communication scenarios are (1) discourses for delivering information as a one-way collaboration, (2) dialogues for exchange of information and relationship building in a two-way collaboration, (3) diatribes for expression of emotions, browbeat, or inspiration as a one-way competition, and (4) debates for winning or convincing in a two-way competition.

Models are an essential instrument in these four scenarios. The communication scenario is sequence of possible communication acts as displayed in Fig. 3. Models are send from a speaker to a hearer based on a speaker-hearer social relationship that supports direct interpretation of the speakers' communication message by the hearers' background, experience, and culture. The model worlds of the speaker and hearer depend on the four driving directives of models: context of the communication partners, community of practice to which the partner belong, scope for origins considered by the partners, and profile (goals, purposes, and functions) of the models to be used.

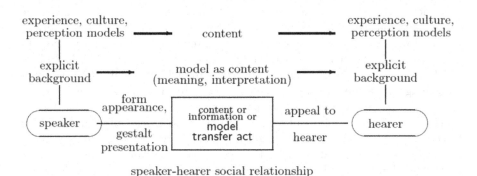

Fig. 3. Dimensions of the communication act for modelling (modified from [20])

The model has several functions in this scenario: (personal/public/group) *recorder of settled or arranged issues, transmitter of information, dialogue service,* and *pre-binding.* Users act in the speaker, hearer, or digest mode.

The communication act is composed of six sub-activities: derive for communication, transfer, receive, recognise and filter against knowledge and experience, understand, and integrate.

We may distinguish two models at the speaker side and six models at the hearer side: speaker's extracted model for transfer, transferred model for both, hearer's received model, hearer's understanding and recognition model, hearer's filtered model, hearer's understood model, and hearer's integration model. These models form some kind of a model ensemble or more formal a model suite (discussed below). Some are extensions or detailing ones; others are zooming ones. Communication is based on some common understanding or at least on transformation of one model to another one. The exchange among all these models can be based on the notion of model capsules (discussed below).

Models for Understanding

Models may be used for understanding the conceptions behind. For instance, conceptualisation is typically shuffled with discovery of phenomena of interest, analysis of main constructs and focus on relevant aspects within the application area. The specification incorporates concepts injected from the application domain.

The function of a model within these scenario is *semantification* or *meaning association* by means of concepts or conceptions. The model becomes enhanced what allows to regard the meaning in the concept.

Models tacitly integrate knowledge and culture of design, of well-forming and well-underpinning of such models and of experience gained so far, e.g. meta-artifacts, pattern and reference models. This experience and knowledge is continuously enhanced during development and after evaluation of constructs.

Models are abstractions and especially truncations of a body of knowledge and information. Humans, sciences, and technology use model within a certain focus and with some analogy. Models can be intentionally wrong on purpose to a certain extend. The Kirchhoff electricity laws can be represented by hydrology models that are not at all entirely consistent models of circuit. The pumping functioning of the human heart can be represented by circuits. We are often concurrently using models that are not consistent in the sense that a holistic model exists which combines all models into one model. Instead we may weaken the consistency property towards a coherence property for model suites. Models are consistent for their sub-models. This recognition is similar to the concept of knowledge islands [17] where sets of formulas are partially consistent for some subsets but not completely consistent. This partial consistency as a specific form of coherence is covered by model suites.

Models are functioning for *elaboration, exploration, detection*, and *acquisition* of tacit knowledge behind the origins which might be products, theories, or engineering activities. They allow to understand what is behind drawn curtain.

Models for Search

Users often face the problem that their mental model and their fact space are insufficient to answer more complex questions [13]. Therefore, they seek information in their environment, e.g. from systems that are available. Information is data that have been shaped into a form that is meaningful and useful for human beings. Information consists of data that are represented in form that is useful and significant for a group of humans. This information search is based on their on the *information need*, i.e. a perceived lack of some information that is desirable or useful. The information is used to derive the current *information demand*, i.e. information that is missing, unknown, necessary for task completion, and directly requested. Is is thus related to the task portfolio under consideration and to the intents.

Search is one of the most common facilities in daily life, engineering, and science. It requires to examine the data and information on hand and to carefully look at or through or into the data and the information.

There is a large variety of information search [6] such as:

1. querying data sets (by providing query expressions in the informed search approach),
2. seeking for information on data (by browsing, understanding and compiling),
3. questing data formally (by providing appropriate search terms during step-wise refinement),
4. ferreting out necessary data (by discovering the information requested by searching out or browsing through the data),
5. searching by associations and drilling down (by appropriate refinement of the search terms),
6. casting about and digging into the data (with a transformation of the query and the data to a common form), and
7. zapping through data sets (by jumping through provided data, e.g., by partially uninformed search).

Search is based on a body of data, information, and knowledge. It is, moreover, based on the background of all models known to the user. A search request or question is formulated according to this background, according to the culture a user has, according to the experience gained so far, and according to the expected answer. This internal background of a search question can be externalised based on the notion of deep model. A deep model [24] is given by the grounding and by the basis of a model. A normal model is rather a surface model, i.e. anything what can be directly captured.

Search is also based on explicit or implicit search habits of a user. Habits are often based on methodologies or experiences acquainted somehow. Language-based search often follows a question-answer form [13], i.e. a question formation includes at least implicitly the form of the expected answer. A question can be considered as a nested quadruple (question content, matter (concepts, situation), user(profile, portfolio), carrier). An answer to such questions that are expected from a system are given by the nested tuple (answer content, solution (characteristics, context, value)).

Search is rather seldom based on a direct question followed by a completely sufficient answer. More often it is a longer search process. There are several kinds of search request broadening and refinement. We may use the function-alisation approach [2] that combines instantiation, context enhancement and refinement. Instantiation is based on parameter assignments. Context enhancement integrates additional question-answer forms what is essentially a model suite composition. Search based on refinement allows an OLAP-like refinement by drill-down, roll-up, dice, slice, and rotate features.

A search model thus incorporates a mould as a distinctive form in which a model is composed. A methodology is a simple mould. A mould can be as complex as it is in manufacturing. Moulds are widely used in science, engineering, and daily life. For instance, mathematicians learn to solve differential equations based on a mould that uses a categorisation of equations and proposes a number of trial-and-error solution forms for solution of the equation. Depending on the culture background of a user that asks a question, we distinguish question-answer forms for linear-active, multi-active, and reactive users. The mould and the deep model form the model matrix as a convenient and effective approach for handling models without repeating details that are accepted anyway. The mould itself is also a macro-model or strategic model. The models derived from search requests are essentially micro-models. If we aggregate these micro-models then we essentially derive meso-models.

The search request can often be mapped to corresponding data structures where the question-answer form is mapped to a series of system-oriented input-output forms.

Models for Analysis

Data analysis, data mining or general analysis combines engineering and (systematic) mathematical problem solving [19]. The model development process combines problem specification and setting with formulation of the analysis tasks by means of macro-models, integration of generic models, selection of the analysis strategy and tactics based on methodology models, models for preparation of the analysis space, and model combination approaches for development of the final model society as the analysis result [14, 15]. The typical process model that governs the analysis process is based on a layering approach, e.g. initial setting, strategy, tactics with generic (or general parameterised models), analysis initialisation, puzzling the analysis results, and final compilation. It is similar to experiment planning in Natural Sciences. The analysis puzzling may follow a number of specific scenarios such as pipe scenarios [18].

Experimental reasoning and empiric analysis are nowadays also branches of computer science. They are crossing two different kinds of reasoning: quantitative reasoning based on data (performance data, behaviouristic data, big data, etc.) and qualitative reasoning based on concepts, conceptions and theories. Quantitative data are used for detection of pattern that can be observed for the given data corpus. Whether these pattern are also valid for other data sets is an open research issue. If the are representative we may call them pre-concepts.

A pattern is a profound, recurring, and consolidated problem-solution pair in a given context where it is useful and will be probably useful to others and in a consistent or coherent format whereas the pattern does not only document 'how' a solution solves a problem but also 'why' it is solved and what are the consequences of this solution.

These pattern are the basis for concept development for quantitative investigations. They become elements of qualitative models. If a pattern can be described by the data set and by an intension then it becomes a (codified) theoretical concept. We may call this modelling approach also conceptional modelling since concepts are extracted and composed from these pattern. Qualitative models can also be based on hypotheses. They can be generalised to some unified holistic theory. These theories can again be used as guidance for further empirical enquiries. This approach has been considered as middle-range theory. In reality, we use, however, models as a mediator between quantitative and qualitative investigations [15]. These mediators are from the side of data interpretational models and from the side of theories either pre-theories or investigative models or guidance models for empiric or experimental investigation. Figure 4 displays the function of an analysis model as a mediating instrument for the combination of empiral science and data exploring science.

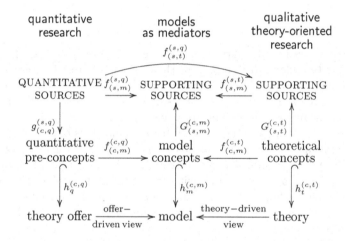

Fig. 4. Models as integrating and mediating instrument for data science

Based on [15], we can now distinguish f_y^x-mappings at the same level, g_y^x-mappings between sources and concept, G_y^x-mappings from concepts to supporting sources, and h_y^x-embedding mappings from concepts to theory offers, models, or theories. Quantitative pre-concepts are indicators or general quantitative properties. Model concepts are already abstractions from those quantitative concepts. Theoretical concepts are elements of a theory that is currently under development. The research task is the harmonisation of the two mappings $f_{(c,m)}^{(c,q)}$ and $f_{(c,m)}^{(c,t)}$. This harmonisation can be based on the mappings for supporting resources $f_{(s,m)}^{(s,q)}$ and $f_{(s,t)}^{(s,q)}$ if some commuting diagram properties are valid for model concepts and the model.

This approach closes the gap between qualitative and quantitative research. Models can be used to render the theory offer. At the same time, models may also render a qualitative theory. The rendering procedures are typically different. A model suite can now be constructed by models for theoretical concepts from one side and by models for quantitative concepts from the other side. In this case, we use models for the quantitative theory offers and for the qualitative theories.

3   Model Conceptions for These Scenarios

It seems that these scenarios require completely different kinds of models. This is however often not the case. We can develop stereotypes which are going to be refined to pattern and later to templates as the basis for model development. We demonstrate for the four scenarios (communication, understanding, search, analysis) how models can be composed in a specific form and which kind of support we need for model-backed collaboration.

Deep Models

A typical model consists of a normal (or surface) sub-model and of deep (implicit, supplanted) sub-models which represent the disciplinary assumptions, the background, and the context. The deep models are the intrinsic components of the model. Conceptualisation might be four-dimensional: sign, social embedding, context, and meaning spaces. The deep model is relatively stable. In science and engineering it forms the disciplinary background. It is often assumed without mentioning it. For instance, database modelling uses the paradigms, postulates, assumptions, commonsense, restrictions, theories, culture, foundations, practices, and languages as carrier within the given thought community and thought style, methodology, pattern, and routines. This background is assumed as being unquestionable given. The normal model mainly represents those origins that are really of interest.

The deep model combines the unchangeable part of a model and is determined by (i) the grounding for modelling (paradigms, postulates, restrictions, theories, culture, foundations, conventions, authorities), (ii) the outer directives (context and community of practice), and (iii) the basis (assumptions, general concept space, practices, language as carrier, thought community and thought style, methodology, pattern, routines, commonsense) of modelling. The deep model can be dependent on mould principles such as the conceptualisation principle [10].

A typical set of deep models are (the models and) foundations behind the origins which are inherited by the models of those origins. Also modelling languages have there specific deep parts. As well as methodologies or more generally moulds of model utilisation stories.

Model Capsules

Model capsules follow a global-as-design approach (see Fig. 5). A model has a number of sub-models that can be used for exchange in collaboration or communication scenarios. A model capsule consists of a main model and exchange sub-models. Model capsules are stored and managed by their owners. Exchange sub-models are either derived from the main model in dependence on the viewpoint, on foci and scales, on scope, on aspects and on purposes of partners or are sub-models provided by partners and transformed according to the main model. A sub-model might be used as an export sub-model (e.g. $A_{4,E}$) that is delivered to the partner on the basis of the import sub-model (e.g. $B_{4,I}$). The sub-models received are typically transformed. We thus use the E(xtract)T(ransform)L(oad) paradigm where extraction and loading is dependent on the language of the sending or receiving model and where transformation allows adaptation of the export sub-model to the import sub-model.

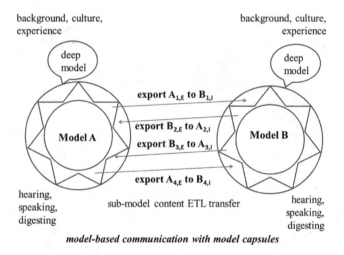

Fig. 5. Exchange on the basis of model capsules with sub-models in model-based ETL-oriented communication scenarios

Model Suites

Most disciplines simultaneously integrate a variety of models or a *society of models*, e.g. [4,12]. The four aspects in Fig. 2 are often given in a separate form as an integrated society of models. Models developed vary in their scopes, aspects and facets they represent and their abstraction.

A typical case are the four aspects that might coexist within a complex model. For instance, models in Egyptology [5][3] can be considered have four

---

[3] The rich body of knowledge resulted in [21] or the encyclopedia with [16].

aspects where each of the aspects has its specific model. The entire model is an integrated combination of (1,2) signs in textual representation and an extending it hieroglyph form (both as representation), (3) interpretation pattern (as the foundation and integration into the thoughts), (4) social determination (as the social aspect), and (5) a context or realisation models into which the model is embedded. The co-design framework for information systems development (integrated design of structuring, functionality, interaction, and distribution) uses four different interrelated and interoperating modelling languages. These modelling languages are at the same level of abstraction and may be combined with additional orientation on usage (as a social component, e.g. represented by storyboards [20]). In this case, the foundational aspect is hidden within the modelling language and within the origins of the models, for instance in the conceptualisation. Following the four aspects in Fig. 2, we derive now models that consider one, two, three, or all four aspects (Fig. 6).

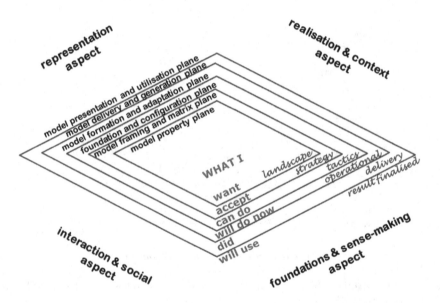

Fig. 6. The four aspect model suite and the corresponding planes for the layers within a model. Activities are governed at each plane by the WHAT I <actually_consider> as main activity.

Figure 6 supports the layers in Fig. 2 by a model suite representation. The model property layer is based on a corresponding plane that essentially considers a model as an adequate and focused, serving the purpose as well as being justified and of acceptable quality. The strategy plane is based on representational features for a useful representation (wording and formulations), on the mould and especially the context, on the systematification, and on an acceptability criterion. At this layer we might already use four different models in a model suite. The tactics plane uses the signification by the given representational features, the integration within the model suite, the enhancement by meaning and thoughts, and the relishes by members of the community of practice. Again we might use different models from the model suite for this plane. At the operational plane we use models that integrate aspects or represent aspects in a separate form, e.g. sense making representations, technologically managed model elements, conceptualisations for foundations, and culture-oriented and stereotyped specific member-oriented models. At the model delivery plane, model suites can be considered to be own models for members of the community of practice which are mastering by tooling, which are corroborated and coherent, and are considered to be appropriate or 'beautiful' for the community. Finally, at the utilisation plane we use naratively represented models, models that are ready for deployment, models that provide a well-understood solution, and models that are inspiring, encouraging, and well-formed.

A *model suite* consists of set of models $\{M_1, ..., M_n\}$, of an association or collaboration schema among the models, of controllers that maintain consistency or coherence of the model suite, of application schemata for explicit maintenance and evolution of the model suite, and of tracers for the establishment of the coherence.

Model suites typically follow a local-as-design paradigm of modelling, i.e. there must not exist a global model which combines all models. In some cases we might however construct the global model as a model that is derived from the models in a model suite. The two approaches to model-based exchange can be combined. A model capsule can be horizontally bound to another capsule within a horizontal model suite or vertically associated to other model capsules. Model capsules are handled locally by members in a team. For instance, model capsules are based on models A and B that use corresponding scientific disciplines and corresponding theories as a part of their background. The models have three derived exchange sub-models that are exported to the other capsule and that are integrated into the model in such a way that the imported sub-model can be reflected by the model of the capsule.

Spaces for Models

Figures 2 and 6 use six planes for detailing models. Each of the planes has it specific quality requirements, support tools, and tasks. At the landscape layer we determine the orientation of the model that should be developed, its problem space, its focus and scope, its integration into the value chain of the application (domain), and its stakeholder from the community of practice with their specific interests and their responsibilities, We rely on mental and codified concepts which are often provided by the world of the origins that a model should properly reflect. The strategic layer adds to this the 'normal way' of development for utilisation of models as methodologies or mould, the embedding into the context and especially the infrastructure, the disciplinary school of thought or more generally the background of the model. The tactics plane embeds the foundations into the modelling process, for instance, by deep model incorporation. It also allows to sketch and to configure the model. The operational plane orients on the formation of the model and the adaptation to the relevant origin(al)s that are going to represented by the model(s). The main issue for the delivery plane is the design of the model(s). The last plane orients utilisation of the model(s) that have been developed. This outer plane might also be structured according to the added value that a model has for the utilisation scenario. Each plane allows to evaluate the model according to quality characteristics used in the sufficiency portfolio.

The model planes have their own workspaces and workplaces which are part of the infrastructure for modelling and utilisation.

4   Model Development

Model Development Story

The modelling story consists of the development story and of the utilisation story. The model development story integrates activities like

1. a selection and construction of an appropriate model according to the function of the model and depending on the task and on the properties we are targeting as well as depending on the context of the intended outcome and thus of the language appropriate for the outcome,
2. a workmanship on the model for detection of additional information about the original and of improved model,
3. an analogy conclusion or other derivations on the model and its relationship to the application world, and
4. a preparation of the model for its use in systems, for future evolution, and for change.

Model utilisation additionally uses assured elementary deployment that includes testing and model detailing and improvement. It may be extended to paradigmatic and systematic recapitulation due to deficiencies from rational and empirical perspectives by the way(s) incommensurability to be resolved. Model deployment also orients on the added value in dependence on the model function in given scenarios. A typical model mould is the mathematics approach to modelling based on (1) exploration of the problem situation, (2) development of an adequate and dependable model, (3) transformation of the first model to a mathematical one that is invariant for the problem formulation and is faithful for the solution inverse mapping to the problem domain, (4) mathematical problem solution, (5) mathematical verification of the solution and validation in the problem domain, and (6) evaluation of the solution in the problem domain [1].

Greenfield Development

Although development from scratch is rather seldom in practice nd daily life we will start with the activities for model development. These activities can be organised in an explorative, iterative, or sequential order in the way depicted in Fig. 6. We can separate activities into[4]:

---

[4] As a generalisation, reconsideration of [11].

(1) **Exploration** of the origin(al)s what results in a well-understood domain-situation and perception models: The origin(al)s will be disassembled into a collection of units. We ensemble (or monstrate) and manifest the insight gained so far in a domain-situation model and develop nominal or perception models for the community of practice. It is based on a plausible model proposition, on a selection of appropriate language and of theories, on generic models, and on commonsense structuring.

(2) **Model amalgamation and adduction** is going to result in a plausible model proposition according to the selected aspects of the four aspects. Amalgamation and adduction are based on an appropriate empirical investigation on origin(al)s, on agreed consensus in the school of thought within the community of practice, on hypothetical reasoning, and on investigative design.

(3) **Final model formulation** results in an adequate and dependable model that will properly function in the given scenarios. We use appropriate depictions for a viable but incomplete model formulation, extend it by corroborated refinements and modifications, and rationally extrapolate the model in dependence on the given ensemble of origin(al)s. In order to guarantee sufficiency of the model, we assess by elementary and prototypical deployment for proper structuring and dependability, within the application domain, within the boundaries of the background, and within the meta-model or mould for model organisation.

A number of moulds can be used for refinement of this development meta-model such as agile or experience-backed methodologies Modelling experience knowledge development might be collected in a later rigor cycle (see design science, for instance, [30]). Model development is an engineering activity and thus tolerates insufficiencies and deficiencies outside the quality requirements. A model must not be true. It must only be sufficient and justified. It can be imperfect.

The result of development can also be a model suite or a model capsule. For instance, information system modelling results in a conceptual structure model, a conceptual functionality model, a logical structure and functionality model, and a physical structure and functionality model. It starts with a business data and process viewpoint model.

Model development can be based on a strictly layered approach in Fig. 7 and on planes in Fig. 6.

Brownfield Development

Modelling by starting from scratch ('greenfield') must be extended by methods for 'brownfield' development that reuses and re-engineers models for legacy systems and within modernisation, evolution, and migration strategies The corresponding model already exists and must be revised. It may also need a revision of its deep sub-model, its basis and grounding, and its ensemble of origin(al)s. All activities used for greenfield development might be reconsidered and revised.

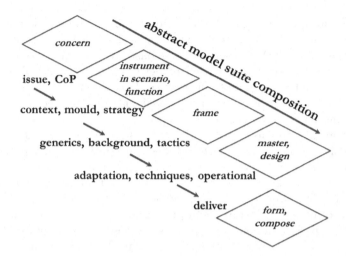

Fig. 7. Model suite development mould for some of the four aspects in Fig. 2

5   Conclusion

Models are widely used and therefore many-facetted, many-functioning, many-dimensional in their deployment, and either macro- or meso- or micro-models. Models are formed according to the function a model has in which scenario. The variety of functions is the reason why models in various scientific and engineering disciplines seem to be different in way that the notions of models cannot be harmonised. We have seen, however, that functions determine to composition of models. Typical composition methods are construction of model suites, model capsules, and inner structuring of models into normal sub-models and deep sub-models. In this paper, we analysed four application areas for modelling: communication, understanding, search, and analysis. We demonstrated how a notion of model is still valid for these four different scenarios. In a similar way other variations of notions of model can be developed. In a university-wide common initiative at Kiel university in a group of more than 40 chairs from almost all faculties, we explored the ingredients of models [28]. The models considered have not only been models used in computer science and computer engineering but also models widely used in archeology, arts, biology, chemistry, ecology, economics, electrotechnics, environmental sciences, farming and agriculture, geosciences, historical sciences, humanities, languages and semiotics, life sciences, mathematics, medicine, ocean sciences, pedagogical science, philosophy, physics, political sciences, social sciences, sociology, and sport science.

The model-being has at least four dimensions which can be grouped into four aspects: *representation* of origins and their specific properties, providing essential *foundations* and thus *sense-making* of origins, relishing and glorifying models as things for *interaction and social collaboration*, and blueprint for *realisation* and constructions within a *context*. This four-aspect consideration directly governs us during introduction of model suites as a model or model capsules. The utilisation scenario and the function of a given model (suite) determine which of the four aspects are represented by a normal model and which aspects are entirely encapsulated in the deep model. The model-being can now be based on demarcation criteria that separate objects and thoughts into models, not-yet-model, and non-models.

Models are embedded into their life, disciplinary, and technical environment, and their culture. They reuse intentionally or edified (or enlightened) existing sub-models, pre-model, reference model, or generic models. A model typically combines an intrinsic sub-model and an extrinsic sub-model. The first sub-model forms the deep model. For instance, database modelling is based on a good number of hidden postulates, paradigms, and assumptions.

The model-being is thus dependent on the scenarios in which models should function properly. We considered here four central scenarios in which models are widely used: communication, understanding, search, and analysis. These four utilisation scenarios can be supported by specific stereotypes of models which model assembling and construction allows a layered mastering of models. The mastering studio has its workspace and its workplace, i.e. in general space for models.

References

1. Berghammer, R., Thalheim, B.: Wissenschaft und Kunst der Modellierung: Modelle, Modellieren, Modellierung, chapter Methodenbasierte mathematische Modellierung mit Relationenalgebren, pages 67–106. De Gryuter, Boston (2015)
2. Bienemann, A.: A generative approach to functionality of interactive information systems. Ph.D. thesis, CAU Kiel, Department of Computer Science (2008)
3. Box, G.E.P.: Science and statistics. J. Am. Stat. Assoc. **71**(356), 791–799 (1976)
4. Coleman, A.: Scientific models as works. Cataloging Classif. Q. Spec. Issue: Works Entities Inf. Retrieval **33**, 3–4 (2006)
5. Deicher, S.: The language of objects. BMBF Project KunstModell in Egyptology (2018). https://www.bmbf.de/files/Kurztexte_SdOIII.pdf
6. Düsterhöft, A., Thalheim, B.: Linguistic based search facilities in snowflake-like database schemes. Data Knowl. Eng. **48**, 177–198 (2004)
7. Embley, D., Thalheim, B. (eds.): The Handbook of Conceptual Modeling: Its Usage and Its Challenges. Springer, Heidelberg (2011)
8. Feyer, T., Thalheim, B.: E/R based scenario modeling for rapid prototyping of web information services. In: Chen, P.P., Embley, D.W., Kouloumdjian, J., Liddle, S.W., Roddick, J.F. (eds.) ER 1999. LNCS, vol. 1727, pp. 253–263. Springer, Heidelberg (1999). https://doi.org/10.1007/3-540-48054-4_21
9. Greefrath, G., Kaiser, G., Blum, W., Borromeo Ferri, R.: Mathematisches Modellieren für Schule und Hochschule, chapter Mathematisches Modellieren - Eine Einführung in theoretische und didaktische Hintergründe, pp. 11–37. Springer, Heidelberg (2013). https://doi.org/10.1007/978-3-658-01580-0
10. Van Griethuysen, J.J.: The Orange report ISO TR9007 (1982–1987) Grandparent of the business rules approach and SBVR part 2 - The seven very fundamental principles, May 2009. https://www.brcommunity.com/articles.php?id=b479. Accessed 21 Sept 2017

11. Halloun, I.A.: Modeling Theory in Science Education. Springer, Berlin (2006). https://doi.org/10.1007/1-4020-2140-2
12. Hunter, P.J., Li, W.W., McCulloch, A.D., Noble, D.: Multiscale modeling: physiome project standards, tools, and databases. IEEE Comput. **39**(11), 48–54 (2006)
13. Jaakkola, H., Thalheim, B.: Supporting culture-aware information search. In: Information Modelling and Knowledge Bases XXVIII, Frontiers in Artificial Intelligence and Applications, vol. 280, pp. 161–181. IOS Press (2017)
14. Kiyoki, Y., Thalheim, B.: Analysis-driven data collection, integration and preparation for visualisation. In: Information Modelling and Knowledge Bases, vol. XXIV, pp. 142–160. IOS Press (2013)
15. Kropp, Y.O., Thalheim, B.: Deep model guided data analysis. In: Kalinichenko, L., Manolopoulos, Y., Malkov, O., Skvortsov, N., Stupnikov, S., Sukhomlin, V. (eds.) DAMDID/RCDL 2017. CCIS, vol. 822, pp. 3–18. Springer, Cham (2018). https://doi.org/10.1007/978-3-319-96553-6_1
16. Liepsner, T.F.: Lexikon der Ägyptologie, volume IV, chapter Modelle, pp. 168–180. Otto Harrassowitz, Wiesbaden (1982)
17. Niskier, C., Maibaum, T., Schwabe, D.: A pluralistic knowledge-based approach to software specification. In: Ghezzi, C., McDermid, J.A. (eds.) ESEC 1989. LNCS, vol. 387, pp. 411–423. Springer, Heidelberg (1989). https://doi.org/10.1007/3-540-51635-2_52
18. Nissen, I.: Wissenschaft und Kunst der Modellierung: Modelle, Modellieren, Modellierung, chapter Hydroakustische Modellierung, pp. 391–406. De Gryuter, Boston (2015)
19. Podkolsin, A.S.: Computer-based modelling of solution processes for mathematical tasks. ZPI at Mech-Mat MGU, Moscov (2001). (in Russian)
20. Schewe, K.-D., Thalheim, B.: Design and Development of Web Information Systems. Springer, Chur (2019). https://doi.org/10.1007/978-3-662-58824-6
21. Teeter, E.: Religion and Ritual in Ancient Egypt. Cambridge University Press, Cambridge (2011)
22. Thalheim, B.: Towards a theory of conceptual modelling. J. Univ. Comput. Sci. **16**(20), 3102–3137 (2010). http://www.jucs.org/jucs_16_20/towards_a_theory_of
23. Thalheim, B.: The conceptual model ≡ an adequate and dependable artifact enhanced by concepts. In: Information Modelling and Knowledge Bases, volume XXV of Frontiers in Artificial Intelligence and Applications, vol. 260, pp. 241–254. IOS Press (2014)
24. Thalheim, B.: Normal models and their modelling matrix. In: Models: Concepts, Theory, Logic, Reasoning, and Semantics, Tributes, pp. 44–72. College Publications (2018)
25. Thalheim, B.: Conceptual modeling foundations: the notion of a model in conceptual modeling. In: Liu, L., Özsu, M.T. (eds.) Encyclopedia of Database Systems. Springer, US (2019). https://doi.org/10.1007/978-1-4614-8265-9_80780
26. Thalheim, B.: Models for communication, understanding, search, and analysis. In: Proceedings of XXI (DAMDID/RCDL 2019), CEUR Workshop Proceedings, vol. 2523, Kazan, Russia, 15–18 October 2019, pp. 19–34 (2019)
27. Thalheim, B., Jaakkola, H.: Models and their functions. In: Proceedings of 29'th EJC, pp. 150–169 Lappeenranta, Finland (2019). LUT, Finland
28. Thalheim, B., Nissen, I. (eds.): Wissenschaft und Kunst der Modellierung: Modelle, Modellieren. Modellierung. De Gruyter, Boston (2015)

29. Thalheim, B., Tropmann-Frick, M.: Wherefore models are used and accepted? The model functions as a quality instrument in utilisation scenarios. In: Comyn-Wattiau, I., du Mouza, C., Prat, N. (eds.) Ingénierie Management des Systèmes d'Information, pp. 131–143. Cépaduès (2016)
30. Wieringa, R.J.: Design Science Methodology for Information Systems and Software Engineering. Springer, Heidelberg (2014). https://doi.org/10.1007/978-3-662-43839-8
31. Wittgenstein, L.: Philosophical Investigations. Basil Blackwell, Oxford (1958)

Applied Ontologies for Managing Graphic Resources in Quantitative Spectroscopy

Nikolai Lavrentev, Alexei Privezentsev, and Alexander Fazliev$^{(\boxtimes)}$ (iD)

Institute of Atmospheric Optics, SB RAS, Tomsk 634055, Russia
{lnick,remake,faz}@iao.ru

Abstract. The report presents the tasks on graphical resources management thoroughly describing applied ontologies of GrafOnto research graphics collection used for solving problems of spectroscopy. Two groups of the tasks on graphical resources management are discussed in the paper. The first group is oriented on the quality analysis of graphical resources and the representation of these results as a graphical resources ontology. The second group of tasks is associated with the automatic class creation for the aforementioned ontology. The problems of ontology modularity and automatic classes' generation are being discussed. Examples of solving reduction problem as well as applied ontologies metrics are presented.

Keywords: Research graphical resources classification · Spectroscopic graphical resources ontology

1 Introduction

In the middle of 2000s the emergence of digital scientific libraries with publications as well as Semantic Web approach oriented on semantic description of information resources induced the work on decomposition of resources into smaller parts that require the creation of semantic annotations oriented on the description of domains and various data representations used in them. Various forms of data representation are always used in scientific publications (text, tables, graphics, symbols (for example, formulas), etc.). On the other side researcher got the facilities for storing and presenting large amounts of information, although published data and information was needed for the control of this information quality. Virtual data centers in various domains appeared in the second half of the 2000-th. These data centers usually contained the published data represented in publications in tabular form. In the end of 2000s publications on scientific graphical resources' systematization started to appear Refs. [1–4]. An example of an approach to creating a collection of graphical resources in High Energy Physics is presented in Ref. [5].

The report presents the results of the final stage of scientific plots' systematization in three disciplines of spectroscopy. At the first stage we formed GrafOnto collection of graphical resources [6–10] describing the results of studies on the problems of a water molecule spectral lines' continuum and on spectral properties of weakly bounded complexes and absorption cross-sections used for the photochemical reactions rates'

© Springer Nature Switzerland AG 2020
A. Elizarov et al. (Eds.): DAMDID/RCDL 2019, CCIS 1223, pp. 82–93, 2020.
https://doi.org/10.1007/978-3-030-51913-1_6

calculation. At the second stage the typification of plots and figures as well as the first version of GrafOnto resources ontology was done (see Refs. [11–15]).

In order to upload new datasets into GrafOnto system and support them one has to solve the tasks on managing graphical resources. These are such tasks as specification of informational resources' structure for spectroscopy problems and analysis of resources' validity, control of data completeness and trust estimation. The decision support system which used in management of the collection GrafOnto is based on ontologies describing the primitive and composite plots and figures. Description of these ontologies is the aim of this report.

This work is founded on the conference paper [14], complemented by the quality analysis of the cited collection plots (see Sects. 3–6). Table 1 from that conference paper is replaced by the table, containing the statistics on GrafOnto collection plots and figures types, CPP + OPP pairs and ontology properties, characterizing virtual cited-original plot pair.

2 GrafOnto Collection of Scientific Graphics

The collection is based on a digital library, containing more than a thousand articles. These articles are dedicated to spectroscopy research such as spectral lines' continuum, weakly bound complexes' properties and spectral functions in near and far ultraviolet range. A distinctive feature of the above problems of spectroscopy is that the major part of published data is represented in a form of plots, figures and images.

In order to create a collection, graphical objects should be manually extracted and converted into a digital form. Software used to upload, storage, view, search and integrate graphical resources into collection is original. At present, the collection contains about 3000 original and 1000 cited primitive plots included into 953 composite plots and 163 composite figures as well as about 4000 primitive plots ready for the upload. The uploaded plots describe properties of 19 molecules, 44 complexes and 66 mixtures. Almost a half of primitive plots characterize properties of a water molecule. Collections' plots are related to dozens of physical quantities (functions) and a dozen of physical quantities (arguments). Table 1 demonstrates the number of primitive plots for different plots and figures types. The number of cited plots is greater than a quarter of all primitive plots, while the cited composite plots comprise 40% of that number. The number of cited plot-specific composite plot pairs is less than a quarter of the whole, therefore this paper is mostly about the problems of forming specific composite plots.

It is worth noting that, at present, only a part of the plots from the publication chosen by experts is uploaded into the collection. Other plots will be processed automatically after the software for machine processing of graphical resources is developed. The collection of plots that has already been created will be used as a data set for training a neural network aimed at automatic recognition of scientific graphics.

Table 1. The statistics of GrafOnto collection primitive plots for different plots and figures types.

Figure type	Plot type	Abbreviation of plots and figures	Restrictions	Quantity
Primitive Figure	Original primitive plot	OPP		3012
	Cited primitive plot	CPP		1160
	Expert primitive plot	EPP		299
	Original composite plot	nOPP	$n > 1$	531
	Cited composite plot	nOPP + mCPP	$m \geq 1, n + m > 1$	430
	Virtual specific composite plot	CCP + OPP		222
	Multipaper composite plot	nOPP + mCPP + kOCiP	$k \geq 1, k + n{+}m \geq 2$	149
Composite Figure		nOPP + mCPP + kOCiP + pCCP	$k + n{+}m + p \geq 2$	163

3   Formation of Virtual Specific Composite Plot

All plots are typified and supplemented with metadata in the GrafOnto collection. The metadata describe data properties, the most interesting among which for us is the "data source type" with unit cardinality. This property can take the values "primary", "cited", or "expert". A primary source (plot) is associated with newly published plots or tables. A cited plot is not primary and is linked to the corresponding primary plot. All other plots in the collection are of the "expert" type, i.e., they need in expert assessment to be assigned to one of the two previous specific types.

Cited primitive plots are loaded into the GrafPlot collection in four steps [15]. The application interfaces for implementation of these steps are shown in Figs. 1 and 2.

Assignment of a "cited" status to a primitive plot. B. Linking to a publication which contains the original plot and selection of the original plot. C. Combining a plot (CCP + OPP) of the cited plot (dots) and the compared original.

In the below procedure for creating a primitive plot, attention is focused on "cited" plots. At the first step, the source type value "cited" is selected. This choice initiates the appearance of the property "bibliographic reference to the original plot" in the data source (see Figs. 1A and 1B). The second step begins with the search and selection of a required bibliographic reference and ends with the provision of a link to and a list of original primitive plots in the corresponding publication, placed in the GrafOnto collection.

The third step consists of selection of an original plot corresponding to the cited plot usually via exhaustion of original plots (see Fig. 1B).

Reference search	Type keyword from reference or type auth	Search of reference using keywords
Chosen reference	*I.V. Ptashnik, K.P. Shine, A.A. Vigasin,* Water vapour self-continuum and water dimers: 1. Analysis of recent work, Journal of Quantitative Spectroscopy and Radiative Transfer, 2011, Volume 112, Issue 8, Pages 1286–1303, DOI: 10.1016/j.jqsrt.2011.01.012. [Annotation]	
Figure No. in publication	110. Burch D. (1982) (296K, 300-1100 cm^{-1}	
Plot title	[toolbar] [45] Burch D. Continuum absorption by H_2O, Air Force Geophysics Laboratory report,AFGL-TR-81-0300, Hanscom AFB, MA, 1982.	

A.

Cited reference search	Type keyword from reference or type auth	Search of reference using keywords
Chosen cited reference	*Burch D.E.,* Continuum absorption by atmospheric H_2O, Report AFGL-TR-81-0300, by Ford Aeronutronic to Air Force Geophys. Lab., Hanscom AFB, Massachusets, **1982**, Pages 46. [Annotation]	

Wavenumber (cm^{-1}): (606.38 - 1343.91), Absorption Coefficient (cm^{2}mol^{-1}atm^{-1})

- 1. Approximation of experimental data (296K, 600-1350 cm^{-1}) [H_2O, 296 K, 612.21985 - 1222.6635 Wavenumber (cm^{-1})]
- 1. Approximation of experimental data (392K, 600-1350 cm^{-1}) [H_2O, 392 K, 670.50815 - 1343.911 Wavenumber (cm^{-1})]
- 1. Approximation of experimental data (430K, 600-1350 cm^{-1}) [H_2O, 430 K, 606.38432 - 844.46399 Wavenumber (cm^{-1})]
- 1. Experiment (296K, 600-1350 cm^{-1}) [H_2O, 296 K, 628.70429 - 1196.9754 Wavenumber (cm^{-1})]
- 1. Experiment (392K, 600-1350 cm^{-1}) [H_2O, 392 K, 696.18365 - 1332.8641 Wavenumber (cm^{-1})]
- 1. Experiment (430K, 600-1350 cm^{-1}) [H_2O, 430 K, 624.71618 - 815.15647 Wavenumber (cm^{-1})]

Wavenumber (cm^{-1}): (2399.33 - 2826.56), Absorption Coefficient (cm^{2}mol^{-1}atm^{-1})

2. Approximated curve (338K, 2400-2829cm^{-1}) [H_2O, 338 K, 2402.0203 - 2659.8418 Wavenumber (cm^{-1})]

Intervals on X axis don't match: (339.76879 - 1069.4798)

2. Approximated curve (384K, 2400-2829cm^{-1}) [H_2O, 384 K, 2400.5113 - 2819.0491 Wavenumber (cm^{-1})]

Intervals on X axis don't match: (339.76879 - 1069.4798)

B.

Fig. 1. Steps of loading a cited primitive plot: (A) reference to the publication [16] where the plot is cited; (B) reference to the publication [17] and a list of plots from the collection linked to this reference.

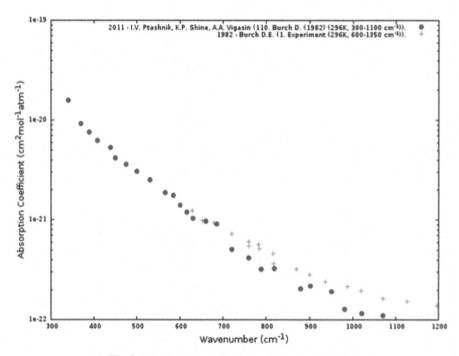

Fig. 2. The last step of loading a cited primitive plot

The absence of an original plot in the collection initiates the analysis of plots and tables from the publication referenced by the data source of the cited plot. If this plot is found, it is incorporated into the GrafOnto collection. Verification for the correspondence of the cited and original plots is the fourth step of the work of this application (Fig. 2).

Figure 1B shows the table with probable cited–original plot pairs and links to their graphical representation.

4 Problems of Matching a Cited Primitive Plot and Its Original

The main problem of linking a cited plot to its original is the definition of the original in the GrafOnto collection. Let us enumerate some tasks the solutions of which are not algorithmized and which should be solved by an expert. The first task is to complete the table of transitions between physical quantities (functions) of different dimensions. The second task is the lack of additional physical quantities in early publications required for the correct transition from one quantity to another. The third task is to search for a publication which contains the original plot if it is published in an old journal, departmental report, dissertation, or conference materials. The fourth task is the search in the case of citing unpublished results with indication of authors or organizations of early publications necessary for the correct transition from one quantity to another.

In the approach suggested, the information contained in the metadata for plots is used to select the original, including coincidence of the intervals of change in coinciding arguments, coincidence of substances, determination of a possibility of recalculating the values of functions identical in sense, but different in dimensions. Figure 1B exemplifies a mismatch of substances under comparison.

In the GrafOnto collection of plots and figures, physical quantities are used as functions (85) or arguments of functions (26). The open set currently contains functions (14 groups) and arguments (8 groups). When grouping arguments, the arguments from different groups are assumed to be unrelated. This assumption is partially used when grouping functions.

5   Managing Graphical Resources in GrafOnto Collection

The principal tasks of graphical resources management are to control resources structure and data quality. An ontology knowledge base accumulating all computer-generated information on collection components is used for making decisions during the management.

Resources structure contains plots of various types, their description, substances, functions and their arguments, physical quantities' units, units table as well as coordinate systems and level of detail of their description, etc. Control of plots and figures validity is based on the analysis of calculated values of paired relations between cited plots and original plots related to them. Such a relation is characterized by a reference to publication, figure number and an identifier of a curve. Note that, at present, the collection of cited plots contains 1160 primitive and 430 composite plots. The ontology describing the present state of the collection resources is presented below.

6   Applied Ontologies of Scientific Plots and Figures in Spectroscopy

Taxonomy of some of the most important artifacts of research publishing [5] includes concepts: figure (composite figure, plot (exclusion area plot, GenericFunctionPlot, histogram), diagram, picture. In our work we defined additional concepts characterized by the methods of acquiring physical quantities (FTP, Cell, etc....) as well as their types (Theoretical, Experimental, Fitting, Asymptotic), slang names of physical quantities, etc. and declared them as subclasses o GenericFunctionPlot class. These definitions are oriented on physical quantities used as plots' axes.

We defined the following hierarchy for forming ontologies in spectroscopy domain. Basic ontology of spectroscopy graphical resources contains three parts and each part is related to one of the three problems of spectroscopy. These problems are the following: problems of continuum absorption, weakly bound complexes as well as the specific task of spectral functions related to photochemical reactions in the atmosphere.

6.1   Basic Ontology and Applied Ontologies of Domain Problems

Basic ontology contains some classes and properties, which are used in applied ontologies of domain problems. In our case, these problems are weakly related to each other and are represented by the following independent modules: graphical resources of continuum absorption, weakly bound complexes and absorption profiles, defining rate of photochemical reaction. Each of these modules is split into three parts: the first part characterizes coordinate systems used in GrafOnto collection, the second one characterizes physical quantities, while the third one characterizes the substances, the properties of which are presented in the collection.

6.2 Main Classes

In ontologies classes define many resources presented in our work in a form of plots and figures from GrafOnto collection as well as in a form of description of their properties. All the classes are explicitly defined in OWL 2 syntax with the use of Manchester syntax for their definition.

Basic Ontology Classes
In the framework of the chosen model the main entities in spectroscopy are substances (*Substance* class) - molecules as well as complexes and mixtures, and methods of acquiring physical quantities' values (*Method*). Graphical representation is related to graphical system entity (*GraphicalSystem*). The components of a graphical system are, for example, the coordinate axes of plots representing physical entities. In GrafOnto each published plot or figure is related to the description of its properties (*Description*, *ResearchPlotDescription* classes). One of such properties is a bibliographic reference to a publication (*Reference*). The *Problem* class contains three individuals (Continuum, Complex and CrossSection) each identifying a problem related to a graphical resource.

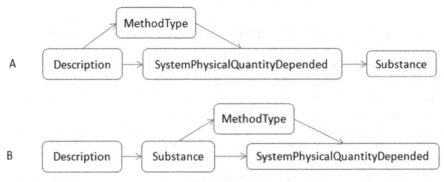

Fig. 3. Word order in the names of A and B groups' classes.

Classes Related to Domain Problems' Ontologies
Domain problems are closely related to the tasks for their solution. GrafOnto collection contains graphical resources related to the problems mentioned in the introductory abstract of this paragraph. Classes of spectroscopy problems' ontologies contain numerous resources and their description. *PhysicalQuantity* class consists of two non-adjacent subclasses named *SystemPhysicalQuantityDepended* and *SystemPhysicalQuantityIndepended*. The first class contains physical quantities the dependency of which on other

physical quantities is presented in plots and figures, while the second one contains physical quantities the dependency of which is presented in plots and figures.

In order to understand the names of ontology classes we have to describe the etymology first. A name may consist of several words. These words correspond to the names of individuals in the corresponding classes *MethodType*, *SystemPhysicalQuantityDepended* and *Substance*. Figure 3 presents examples of schemes for creating subclasses names in A classes (*Physical quantity and related substances*) and B classes (*Substance and related physical quantities*) presented in simplified syntax.

For example, a class named *Description_Experimental_Absorption_Coefficient__cm2mol_1atm_1_* contains all the descriptions of measured absorption coefficients with $cm^2 molecule^{-1} atm^{-1}$ dimension for a series of substances being a subclass of *Physical quantity and related substances*. The third group of classes related to subclasses of *GraphicalSystem* class is not presented in this work.

6.3 Main Properties

Comments for all the properties used in natural language are presented in OWL 2 ontologies code. Here we present a simplified classification of some properties related to physical quantities and descriptions of plots and figures. Description of properties related to Description and CoordinateSystem classes as well as to Temperature and Pressure quantitative characteristics is omitted.

Table 2 lists ontology properties defining their domains and ranges. The last column of the table shows abbreviations of properties used in the scheme of individual presented in Fig. 2 [14].

Qualitative properties characterized physical quantities are *hasOriginType*, *hasSourceType* and *hasMethodType*. The values of *hasOriginType* property indicate the origin of dataset related to the plot: it should be original and should be obtained by digitizing the curve of a primitive plot. The values of *hasSourceType* property can describe primary data, i.e. the data obtained by the authors of the publication as well as the previously published curves (i.e. cited) and commonly known curves (i.e. expert). The values of *hasMethodType* property characterize qualitative acquisition of datasets of primitive plot: Theoretical is a calculation using physical or mathematical model, Experimental is measurement, Fitting is a continuous curve creation using the method of fitting to experimental values.

The relations between plots and figures are defined by 5 properties (*has{OPPD, CPPD, OCPD, CCPD, MCPD}*). These five mereological properties describe composition of composite plots (*OPPD, CPPD*) and figures (*OCPD, CCPD, MCPD*). The value of the *isCitationOf* property used in the description of cited primitive plot is the corresponding original plot. This property defines the descriptions that contain datasets with closely related values.

Table 2. Main properties of base ontology and spectroscopy problems ontology

Domain	Property	Range
Primitive Plot Description (PPD)		
Domain	Property	Range
Description	hasReference	Reference
PrimitivePlotDescription	hasSubstance	Substance
PrimitivePlotDescription	hasSourceType	{Primary, Expert, Cited}
PrimitivePlotDescription	hasOriginType	{Digitized, Original}
PrimitivePlotDescription	hasCurveType	{Line, Point}
PrimitivePlotDescription	hasCS	CoordinateSystem
CitedPrimitivePlotDescription	hasCitedReference	Reference
CoordinateSystem	hasCSType	{2D-Decartes}
CoordinateSystem	hasX-axis	X-axis
CoordinateSystem	hasY-axis	Y-axis
Y-axis	hasMethod	Method
Y-axis	hasMethodType	{Theory, Experiment, Fitting}
Y-axis	hasPY-axis	PubPhysQuanDepended
Y-axis	hasSY-axis	SysPhysQuanDepended
X-axis or Y-axis	hasAxisScale	{Linear, Logarithmic}
X-axis	hasPX-axis	PubPhysQuanIndep
X-axis	hasSX-axis	SysPhysQuanIndep
PrimitivePlotDescription	hasTemperature	float
PrimitivePlotDescription	hasPressure	float
PrimitivePlotDescription	hasSystemFigureNumber	integer
ResearchFigureDescription	hasOriginalImageOfPlot	URI
ResearchFigureDescription	hasOriginalPlotInformation	URI
FigureDescription	hasFigureCaption	string
FigureDescription	isPartOfFigureNumber	integer
ResearchFigureDescription	hasNumberOf Points	integer
ResearchFigureDescription	hasPlotCaption	string
Primitive Plot Description		
CitedPrimitivePlotDescription	isCitationOf	OriginalPrimitivePlotDescription
OriginalPrimitivePlotDescription	hasCitation	CitedPrimitivePlotDescription
Original Composite Plot Description (OCPD)		
OriginalCompositePlotDescription	hasOPPD	OriginalPrimitivePlotDescription
Cited Composite Plot Description (CCPD)		
CitedCompositePlotDescription	hasCPPD	PrimitivePlotDescription
Composite Figure Description (CFD)		
CompositeFigureDescription	hasOCPD	OriginalCompositePlotDescription
CompositeFigureDescription	hasCCPD	CitedCompositePlotDescription
CompositeFigureDescription	hasMCPD	MultipaperCompositePlotDescription

6.4 Main Types of Individuals

Being equivalents of figures and plots from published graphical resources on the above problems images generated in GrafOnto system are related to the description of their metadata making the most significant part of ontology individuals included in A-box. Typification of figures and plots given in Ref. [12] is defined by the property values. Abbreviation of corresponding values is used in the names of such individuals (for example, OCP - Original Composite Plot). Figure 2 in [14] illustrates the structure of one of such plot types, i.e. original primitive plot. Ovals stand for ontology individuals, rectangles stand for literals and directed arcs stand for objective (OP) and determined (datatype - DTP) properties. Cited primitive plot have a similar structure with an addition of observations with hasPrototype, hasChild and hasParent properties. Special cases of individuals characterizing properties of coordinate system and its axes are shown in the lower part of Fig. 2 (see Ref. [14]). A series of individuals are related to the classes defined by enumeration of its individuals.

Ontologies metrics are used for comparing ontologies of different parts of a domain or of different domains, characterizing quantitative and qualitative peculiarities of ontological description. In OWL ontologies the number of object properties characterizes the number of paired relations between individuals. Some of these individuals may have quantitative estimation. The estimated relations are described by certain (datatype) properties.

Table 3. Ontology metrics, characterizing the graphical resources collection

	Continuum	Complex	Cross Section
Metrics			
Axiom	69730	28982	10605
Logical axiom count	56898	23538	8566
Declaration axiom count	10213	4432	1682
Class count	34	34	32
Object properties count	24	24	24
Datatype properties count	14	14	14
Individual count	10290	4581	1632
DL expressivity	ALCO(D)	ALCO(D)	ALCO(D)
Individual axioms			
Class assertion	442	278	198
Object properties assertion	37467	15553	5582
Datatype properties assertion	18878	7596	2675
Annotation axioms			
Annotation assertions	2619	1012	355

Table 3 contains metrics for applied ontologies of three spectroscopy problems as well as the unification of these ontologies (Σ Ontology). As for GrafOnto resources collection the equality of numbers characterizing the number of properties, their domains and ranges means that they are characterized by identical properties. However, the difference in classes' numbers indicates the use of a greater number of spectral functions in Continuum problem in comparison with Complex and Cross Section problems. As individuals of one and the same group of types are used applied ontologies we may conclude that ontology on Continuum problem describe the highest number of primitive plots.

Metrics comparison of Σ Ontology with ontologies of tabular information resources Ref. [14] reveals that in our work on graphical resources ontology we managed to significantly increase the number of classes in one year. It clearly indicates that Σ Ontology contains the highest number of obvious answers on typical user requests.

7 Conclusion

The report presents applied ontologies of scientific plots and figures used for managing graphical resources in three problems of spectroscopy. One of the basic GrafOnto collection management problems is described: the problem of building virtual composite plots "cited-original plot". We propose to use an OWL-ontology to control the quality of distributed scientific graphics collection. Ontologies describe a collection of plots and figures published in the period from 1918 till 2019. Ontologies are created for managing structure of collection resources as well as for making decisions on such tasks as development, storage and systematization of plots and figures for solving such problems as continuum absorption and research of properties of weakly related complexes and cross sections absorption. Ontology as well as its individuals and classes are automatically generated with the enlargement of the collection.

Acknowledgements. The authors are grateful to Profs. I.V. Ptashnik and N.N. Filippov for providing original data pertaining to figures in their publications. The work was supported by the Basic Research Program no. AAAA-A17-117021310148-7.

References

1. Halpin, H., Presutti, V.: An ontology of resources for linked data. Linked Data on the Web, Madrid, Spain. ACM 978-1-60558-487-4/09/04 (2009)
2. Thorsen, H., Pattuelli, C.M.: Ontologies in the time of linked data. In: Smiraglia, R.P. (ed.) Proceedings from North American Symposium on Knowledge Organization, vol. 5, pp. 1–15 (2015)
3. Niknam, M., Kemke, C.: Modeling Shapes and Graphics Concepts in an Ontology. https://pdfs.semanticscholar.org/c20b/3b819ce253715bbfa9c2151a10ea87f718e4.pdf
4. Kalogerakis, E., Christodoulakis, S., Moumoutzis, N.: Coupling ontologies with graphics content for knowledge driven visualization. https://people.cs.umass.edu/~kalo/papers/graphicsOntologies/graphicsOntologies.pdf

5. Praczyk, P.A.: Management of Scientific Images: an approach to the extraction, annotation and retrieval of figures in the field of high energy physics. Thesis Doctoral, Universidad de Zaragoza (2013). ISSN 2254-7606
6. Voronina, Yu.V., Lavrentiev, N.A., Privezentzev, A.I., Fazliev, A.Z.: Collection of published plots on water vapor absorption cross sections. In: Proceedings of the 24th International Symposium on Atmospheric and Ocean Optics: Atmospheric Physics, vol. 10833. SPIE (2018). https://doi.org/10.1117/12.2504586
7. Lavrentiev, N.A., Rodimova, O.B., Fazliev, A.Z., Vigasin, A.A.: Systematization of published research plots in spectroscopy of weakly bounded complexes of molecular oxygen and nitrogen. In: Proceedings of the 24th International Symposium on Atmospheric and Ocean Optics: Atmospheric Physics, vol. 10833. SPIE (2018). https://doi.org/10.1117/12.2504327
8. Lavrentiev, N.A., Rodimova, O.B., Fazliev, A.Z.: Systematization of published scientific graphics characterizing the water vapor continuum absorption: I publications of 1898–1980. In: Proceedings of the 24th International Symposium on Atmospheric and Ocean Optics: Atmospheric Physics, vol. 10833. SPIE (2018). https://doi.org/10.1117/12.2504325
9. Lavrentiev, N.A., Rodimova, O.B., Fazliev, A.Z., Vigasin, A.A.: Systematization of published research graphics characterizing weakly bound molecular complexes with carbon dioxide. In: Proceedings of the 23rd International Symposium on Atmospheric and Ocean Optics: Atmospheric Physics, vol. 104660E. SPIE (2017). https://doi.org/10.1117/12.2289932
10. Lavrentiev, N.A., Rodimova, O.B., Fazliev, A.Z.: Systematization of graphically plotted published spectral functions of weakly bound water complexes. In: Proceedings of the 22nd International Symposium on Atmospheric and Ocean Optics: Atmospheric Physics, vol. 10035. SPIE (2016). https://doi.org/10.1117/12.2249159
11. Lavrentiev, N.A., Privezentsev, A.I., Fazliev, A.Z.: Tabular and graphic resources in quantitative spectroscopy. In: Manolopoulos, Y., Stupnikov, S. (eds.) DAMDID/RCDL 2018. CCIS, vol. 1003, pp. 55–69. Springer, Cham (2019). https://doi.org/10.1007/978-3-030-23584-0_4
12. Lavrentiev, N.A., Privezentsev, A.I., Fazliev, A.Z.: Systematization of tabular and graphical resources in quantitative spectroscopy. In: Kalinichenko, L., Manolopoulos, Y., Stupnikov, S., Skvortsov, N., Sukhomlin, V. (eds.) CEUR Workshop Proceedings, Selected Papers of the XX International Conference on Data Analytics and Management in Data Intensive Domains, vol. 2277, pp. 25–32 (2018)
13. Lavrentiev, N.A., Privezentsev, A.I., Fazliev, A.Z.: Applied ontology of molecule spectroscopy scientific plots. In: Proceedings of the Conference Knowledge, Ontologies, Theories, vol. 2, pp. 36–40. DigitPro (2017)
14. Lavrentiev, N.A., Privezentsev, A.I., Fazliev, A.Z.: Applied ontologies for managing graphic resources in spectroscopy. In: Elizarov, A., Novikov, B., Stupnikov, S. (eds.) CEUR Workshop Proceedings, Selected Papers of the XX International Conference on Data Analytics and Management in Data Intensive Domains, vol. 2523, pp. 107–116 (2019)
15. Akhlestin, A.Yu., Lavrentiev, N.A., Rodimova, O.B., Fazliev, A.Z.: The continuum absorption: trust assessment of published graphical information. In: Proceedings of the SPIE 11208, 25th International Symposium on Atmospheric and Ocean Optics: Atmospheric Physics, p. 112080P (2019). https://doi.org/10.1117/12.2541741
16. Ptashnik, I.V., Shine, K.P., Vigasin, A.A.: Water vapour self-continuum and water dimers: 1. Analysis of recent work. J. Quant. Spectrosc. Radiat. Transfer **112**(8), 1286–1303 (2011). https://doi.org/10.1016/j.jqsrt.2011.01.012
17. Burch, D.E.: Continuum absorption by atmospheric HO. Report AFGL-TR-81-0300, p. 46, Ford Aeronutronic to Air Force Geophys. Lab., Hanscom AFB, Massachusets (1982)

Data Analysis in Astronomy

Realization of Different Techniques for Anomaly Detection in Astronomical Databases

Konstantin Malanchev[1,2]($\boxtimes$), Vladimir Korolev[3,4], Matwey Kornilov[1,2], Emille E. O. Ishida[5], Anastasia Malancheva[6], Florian Mondon[5], Maria Pruzhinskaya[1], Sreevarsha Sreejith[5], and Alina Volnova[7]

[1] Sternberg Astronomical Institute, Lomonosov Moscow State University, Universitetsky pr. 13, Moscow 119234, Russia
malanchev@physics.msu.ru
[2] National Research University Higher School of Economics, 21/4 Staraya Basmannaya Ulitsa, Moscow 105066, Russia
[3] Central Aerohydrodynamic Institute, 1 Zhukovsky Street, Zhukovsky, Moscow Region 140180, Russia
[4] Moscow Institute of Physics and Technology, 9 Institutskiy per., Dolgoprudny, Moscow Region 141701, Russia
[5] Université Clermont Auvergne, CNRS/IN2P3, LPC, 63000 Clermont-Ferrand, France
[6] Cinimex, Bolshaya Tatarskaya Street 35 bld. 3, Moscow 115184, Russia
[7] Space Research Institute of the Russian Academy of Sciences (IKI), 84/32 Profsoyuznaya Street, Moscow 117997, Russia

Abstract. In this work we address the problem of anomaly detection in large astronomical databases by machine learning methods. The importance of such study is justified by the existence of a large amount of astronomical data that can not be processed only by human resource. We evaluated five anomaly detection algorithms to find anomalies in the light curve data of the Open Supernova Catalog. Comparison of the algorithms revealed that expert supervised active anomaly detection method shows the best performance, while among purely unsupervised techniques Gaussian mixture model and one-class support vector machine methods outperform isolation forest and local outlier factor methods.

Keywords: Machine learning · Isolation forest · Active anomaly detection · Local outlier factor · One-class SVM · Supernovae

1 Introduction

Modern astronomy produces a huge amount of data obtained by different dedicated surveys and experiments. Terabytes of data require careful processing to extract valuable information. The information about our universe is collected

A. Elizarov et al. (Eds.): DAMDID/RCDL 2019, CCIS 1223, pp. 97–107, 2020.
https://doi.org/10.1007/978-3-030-51913-1_7

daily in every domain of electromagnetic spectrum: in high energy range (Spectr-RG[1]), optics (SDSS[2], Gaia[3]), and radio (EHT[4]), as well as in neutrino astrophysics (IceCube[5]) and gravitational waves (LIGO/Virgo[6]).

In this new paradigm of the exponential growth of astronomical data volume the use of machine learning (ML) methods becomes inevitable in this field [2]. Most of the ML efforts in astronomy are concentrated in classification (e.g., [14]) and regression (e.g., [4]) tasks. Another important problem of the automated data analysis is a search for unusual data samples that correspond to abnormal astronomical objects, e.g. objects of some rare or even novel class. This problem is called the anomaly detection.

Astronomical anomaly detection is the field where ML methods may be used quite effectively taking into account the enormous amount of data that has been gathered, however, they have not been fully implemented yet. Barring a few exceptions, most of the previous studies may be divided into only two different trends: clustering [29] and subspace analysis [12] methods. More recently, random forest algorithms have been used extensively by themselves [3] or in hybrid statistical analysis [25]. It seems to be the most attractive to search for anomalies in synoptic transient surveys since the fast scanning of large field of sky increases the probability to detect new or rare type of objects. That is why the anomaly detection problem is facing all future major sky surveys such as Legacy Survey of Space and Time (LSST[7]). However, it is important to understand that there are many anomaly detection algorithms and the choice of the particular one depends on the concrete task.

In this article we addressed the problem of the effectiveness of certain anomaly detection algorithms as applied to supernovae (SNe) data, since SNe are among the most numerous objects discovered in astronomy. Supernovae surveys detect hundreds of SNe candidates per year, but the lack of spectroscopic information makes the processing algorithms to classify discovered SNe basing on secondary features (proximity to the galaxy, monotonous flux changing with time, absolute magnitude, etc). Anomaly detection may solve two problems: (a) minimize the contamination of non-SNe in large supernova databases, the presence of which is a direct consequence of the lack of spectroscopic support, and (b) find inside the SNe data rare or new classes of objects with unusual properties. To demonstrate that, we apply five different anomaly detection algorithms: Gaussian mixture model (GMM), local outlier factor (LOF), one-class support vector machine (SVM), isolation forest, and active anomaly detection (AAD), to the real photometrical data from the Open Supernova Catalog (OSC[8], [11]).

---

[1] http://srg.iki.rssi.ru/.

[2] https://www.sdss.org/.

[3] https://sci.esa.int/web/gaia.

[4] https://eventhorizontelescope.org/.

[5] https://icecube.wisc.edu/.

[6] https://www.ligo.org/.

[7] https://www.lsst.org/.

[8] https://sne.space/.

The OSC and feature extraction are described in Sect. 2. In Sect. 3 we briefly describe the anomaly detection methods used. In Sect. 4 we present the results of our search of anomalies in the OSC and compare the efficiency of anomaly detection algorithms.

The results presented in this paper are partly based on the publications [15, 21, 27]. In [21] we focused on anomaly detection in the OSC with use of isolation forest algorithm only. Here, we have shown for the first time the comparison between the different anomaly detection algorithms as applied to the same data set as in [21].

2 Data Preprocessing

2.1 Open Supernova Catalog

The Open Supernova Catalog [11] is constructed by combining many publicly available data sources. It includes different catalogs and surveys as well as information from individual studies. It represents an open repository for supernova metadata, light curves, and spectra in an easily downloadable format. This catalog also includes some contamination from non-SN objects. It contains data for more than 55000 SNe candidates among which ~13000 objects have >10 photometric observations and for ~7500 spectra are available.

The catalog stores the light curves (LCs) data in different magnitude systems. Since we need a homogeneous data sample, we extracted only the LCs in BRI, gri, and $g'r'i'$ filters. We assume, that $g'r'i'$ filters are close enough to gri filters to consider them as the same filters. We also transform BRI magnitudes to gri using the Lupton's photometrical equations[9]. We also require a minimum of three photometric points in each filter with 3-day binning.

2.2 Light Curves Approximation

Traditionally, ML algorithms require a homogeneous input data matrix which, unfortunately, is not the case with supernovae. A commonly used technique to transform unevenly distributed data into a uniform grid is to approximate them with Gaussian processes (GP; [28]). Usually, each light curve is approximated by GP independently. However, in this study we use a Multivariate Gaussian Process[10] approximation. For each object it takes into account the correlation between light curves in different bands, approximating the data by GP in all filters in a one global fit (for details see Kornilov et al. 2020, in prep.). As an approximation range we chose $[-20; +100]$ days. We also extrapolated the GP approximation to fill this range if needed. With this technique we can reconstruct the missing parts of LC from its behavior in other filters.

Once the Multivariate Gaussian Process approximation was done, we visually inspected the resulting light curves. Those SNe with unsatisfactory approximation were removed from the sample (mainly the objects with bad photometric

---

[9] http://www.sdss3.org/dr8/algorithms/sdssUBVRITransform.php.
[10] http://gp.snad.space.

quality). Since each object has its own flux scale due to the different origin and different distance, we normalized the flux vector by its maximum value. Based on the results of this approximation, for each object we extracted the kernel parameters, the log-likelihood of the fit, LC maximum and normalized photometry in the range of $[-20, +100]$ days with 1-day interval relative to the maximum. These values were used as features for the anomaly detection algorithms. Our final sample consists of 1999 objects, $\sim$30% of which have at least one spectrum in the OSC. Less than 5% of our sample have <20 photometric points in all three filters.

3   Anomaly Detection

To the data set of approximated light curves of $\sim$2000 objects (see Sect. 2) we applied five different anomaly detection algorithms that described below.

3.1   Gaussian Mixture Model

The idea of this method is to present the data set as a random sample of a mixture of multiple multivariate Gaussian distributions. The data set is fitted to the mixture with the maximum likelihood and then the samples with the smallest probability values are interpreted as outliers (see, e.g., [32] for a description of probability driven anomaly detection). The only hyper-parameter of the Gaussian mixture model is a number of mixture components. We found that eight components give the best performance in our case.

3.2   Local Outlier Factor

Local outlier factor [6] is a name of several methods that find outliers as local deviations of density. We used LOF method based on famous k-nearest neighbours algorithm, which gives a measure of local density of a sample. We found that number of neighbours $k = 40$ gives the best performance in case of our data set.

3.3   One-Class Support Vector Machine

One-class support vector machine [23,31] is based on isolation of outliers in a higher dimensional space given by some kernel function. We used radial-basis function as a kernel function in our analysis.

3.4   Isolation Forest

Isolation forest [18,19] is built on an ensemble of random isolation trees. Each isolation tree is a space-partitioning tree similar to the widely known k-dimensional tree (k-d tree; [5]). However, in contrast to the k-d tree, a space coordinate (a feature) and a split value of an isolation tree are selected at random for every node of the isolation tree. The tree is built until each object of a sample is isolated in a separate leaf – the shorter path corresponds to a higher anomaly score. For each object, the measure of its normality is the arithmetic average of the depths of the leaves into which it is isolated.

Details of isolation forest implementation for the problem of anomaly detection in Open Supernova Catalog can be found in a separate paper [27].

3.5   Active Anomaly Detection

Active anomaly detection [9,10] is a technique that combines some anomaly detection algorithm with an expert feedback. We have used an AAD method based on the same isolation forest as in Sect. 3.4. In this case each leaf is assigned by some weight and anomaly score equation is modified to include the weights. On the first step of AAD algorithm, all the leafs are assigned with the equal weights, the most abnormal object is given to an expert, and the expert decides if this object is a true anomaly or not. If the expert says no, then the leafs are reweighed to decrease significance of the leafs containing the object, and anomaly scores of the objects are updated correspondingly. If the export says yes, the weights are not affected. Then, the algorithm yields an object with the second largest anomaly score and so on until the expert's working time budget runs out.

Our previous study showed that this method provides better results in comparison with ordinary isolation forest [15].

3.6   Random Sampling

To show that all the methods described above yield significant amount of true anomalies, we implemented a simple algorithm that returns objects from the data set one by one without replacement. The performance of such random sampling algorithm is treated as a baseline for all other algorithms.

4   Results

We set the outlier contamination level to 2% for each anomaly detection algorithm. When the list of outliers has been obtained for each algorithm, we inspected visually their LCs and analyzed them using publicly available information. Basing on this analysis, we decided whether the object is an anomaly or not. As an anomaly we defined an object that is not a SN, i.e. belongs to other kind of astrophysical sources; represents some rare supernova subtypes, i.e. peculiar SNe, superluminous supernovae; artefacts in the original data such as misprints in the OSC; and the objects of unknown nature.

Summarizing the outputs of all algorithms, we found several peculiar supernovae of different types, e.g., low-luminosity IIP supernova 2013am (Fig. 1), Ia-91bg supernova 2006mr (Fig. 2), Ibn supernova PS1-12sk (Fig. 3); superluminous SNe, e.g., SN1000+0216 (Fig. 4), SN2213-1745, PTF10aagc, few cases of misclassifications of SNe as active galactic nuclei, stars, binary microlensing event Gaia16aye (Fig. 5).

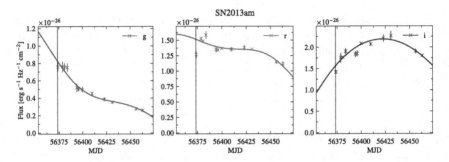

Fig. 1. Light curves in *gri* filters of low-luminosity IIP supernova 2013am [17,24]. Solid lines are the results of our approximation by Multivariate Gaussian Process. The vertical line denotes the moment of maximum in *r* filter.

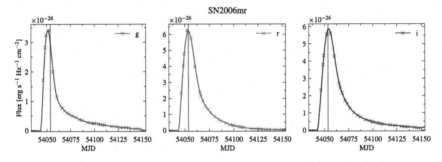

Fig. 2. Light curves in *gri* filters of Ia-91bg supernova 2006mr [7]. Solid lines are the results of our approximation by Multivariate Gaussian Process. The vertical line denotes the moment of maximum in *r* filter.

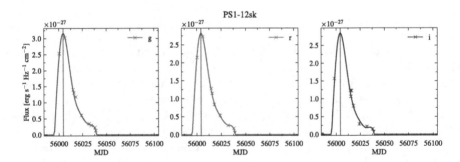

Fig. 3. Light curves in *gri* filters of Ibn supernova PS1-12sk [30]. Solid lines are the results of our approximation by Multivariate Gaussian Process. The vertical line denotes the moment of maximum in *r* filter.

4.1   Comparison of Performance of Anomaly Detection Methods

For each method we obtained a list of anomalies. Contamination levels were set to 2% (40 objects). For the case of active anomaly detection method (Sect. 3.5) the expert provided feedback to the algorithm and examined first 40 objects one by one. For all other methods the expert examined 2% of the objects with the highest anomaly score.

Figure 6 shows how the percentage of true anomalies depends on the number of object examined by the expert. It is clear that any valuable anomaly detection algorithm performs better than random sampling. Active anomaly detection method shows the best performance by rising up to 27% of true anomalies outperforming isolation forest, which AAD is based on, by a factor of two. Other valuable methods show almost the same maximum fraction of anomalies, while one-class support vector machine found the first true anomaly, and local outlier factor method shows the steepest rise. Looking at the fraction of anomalies for 40 candidates we can deduce that active anomaly detection outperforms all pure unsupervised anomaly detection algorithms. Also, we can say with a certain

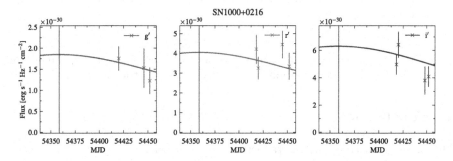

Fig. 4. Light curves in $g'r'i'$ filters of superluminous supernova 1000+0216 [8]. Solid lines are the results of our approximation by Multivariate Gaussian Process. The vertical line denotes the moment of maximum in r' filter.

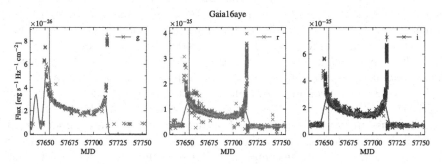

Fig. 5. Light curves in gri filters of binary microlensing event Gaia16aye [1,34]. Solid lines are the results of our approximation by Multivariate Gaussian Process. The vertical line denotes the moment of maximum in r filter.

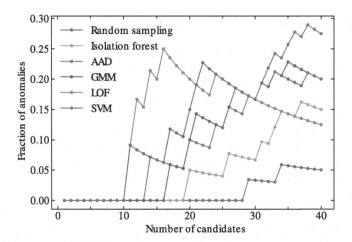

Fig. 6. Fraction of anomalies as a function the total number of candidates (outliers) scrutinised by the expert. The plot shows results obtained with random sampling (blue), isolation forest (orange), active anomaly detection (green), Gaussian mixture model (red), local outlier factor (violet) and one-class support vector machine (brown) algorithms. (Color figure online)

amount of confidence, that Gaussian mixture model and one-class SVM methods perform better than isolation forest and local outlier factor methods.

5   Conclusions

The amount of astronomical data increases dramatically with time and is already beyond human capabilities. The astronomical community already has dozens of thousands of SN candidates, and LSST survey will discover over ten million supernovae in the forthcoming decade [20]. Only a small fraction of them will receive a spectroscopic confirmation. This motivates a considerable effort in photometric classification of supernovae by types using machine learning algorithms. There is, however, another aspect of the problem: any large photometric SN database would suffer from the non-SN contamination (novae, kilonovae, GRB afterglows, AGNs, etc.). Moreover, the database will inevitably contain the astronomical objects with unusual physical properties – anomalies.

In this study, we examined the effectiveness of different anomaly detection algorithms as applied to the Open Supernova Catalog [11]. Each algorithm, namely Gaussian mixture model, local outlier factor, one-class support vector machine, isolation forest, and active anomaly detection, provided 40 outliers that correspond to 2% of the sample size. These outliers were subjected to the careful astrophysical analysis since not necessary all of them are real anomalies. Among found anomalies there are superluminous supernovae, non-classical Type Ia supernovae, unusual Type II supernovae, active galactic nuclei and one binary microlensing event.

The comparison of anomaly detection algorithms revealed that the AAD algorithm outperforms the others. Among the pure unsupervised algorithms the highest fraction of anomalies showed by GMM and SVM. Search for the most performant anomaly detection methods is important in the context of forth-coming large sky surveys, like LSST, to make data processing and analysis efficient and expeditious.

Acknowledgements. M. Kornilov, V. Korolev, K. Malanchev, A. Volnova and M. Pruzhinskaya are supported by RFBR grant according to the research project 20-02-00779 for preparing the Open Supernova Catalog data, anomaly analysis and evaluation of different anomaly detection algorithms. E.E.O. Ishida and Sreevarsha Sreejith acknowledge support from CNRS 2017 MOMENTUM grant. A. Volnova acknowledges support from RSF grant 18-12-00522 for analysis of interpolated LCs. We used the equipment funded by the Lomonosov Moscow State University Program of Development. The authors acknowledge the support from the Program of Development of M.V. Lomonosov Moscow State University (Leading Scientific School "Physics of stars, relativistic objects and galaxies"). This research has made use of NASA's Astrophysics Data System Bibliographic Services and following Python software packages: NUMPY [33], MATPLOTLIB [13], SCIPY [16], PANDAS [22], and SCIKIT-LEARN [26].

References

1. Bakis, V., et al.: Gaia16aye: a flaring object of uncertain nature in Cygnus. The Astronomer's Telegram 9376, August 2016
2. Ball, N.M., Brunner, R.J.: Data mining and machine learning in astronomy. Int. J. Mod. Phys. D **19**, 1049–1106 (2010). https://doi.org/10.1142/S0218271810017160
3. Baron, D., Poznanski, D.: The weirdest SDSS galaxies: results from an outlier detection algorithm. MNRAS **465**, 4530–4555 (2017). https://doi.org/10.1093/mnras/stw3021
4. Beck, R., et al.: On the realistic validation of photometric redshifts. MNRAS **468**, 4323–4339 (2017). https://doi.org/10.1093/mnras/stx687
5. Bentley, J.L.: Multidimensional binary search trees used for associative searching. Commun. ACM **18**(9), 509–517 (1975). https://doi.org/10.1145/361002.361007
6. Breunig, M.M., Kriegel, H.P., Ng, R.T., Sander, J.: LOF: identifying density-based local outliers. In: Proceedings of the 2000 ACM SIGMOD International Conference on Management of Data, pp. 93–104 (2000)
7. Contreras, C., et al.: The Carnegie Supernova Project: first photometry data release of low-redshift type ia supernovae. AJ **139**, 519–539 (2010). https://doi.org/10.1088/0004-6256/139/2/519
8. Cooke, J., et al.: Superluminous supernovae at redshifts of 2.05 and 3.90. Nature **491**, 228–231 (2012). https://doi.org/10.1038/nature11521
9. Das, S., Rakibul Islam, M., Kannappan Jayakodi, N., Rao Doppa, J.: Active Anomaly Detection via Ensembles. arXiv e-prints arXiv:1809.06477, September 2018
10. Das, S., Wong, W.K., Fern, A., Dietterich, T.G., Amran Siddiqui, M.: Incorporating Feedback into Tree-based Anomaly Detection. arXiv e-prints arXiv:1708.09441, August 2017
11. Guillochon, J., Parrent, J., Kelley, L.Z., Margutti, R.: An open catalog for supernova data. APJ **835**, 64 (2017). https://doi.org/10.3847/1538-4357/835/1/64

12. Henrion, M., Hand, D.J., Gandy, A., Mortlock, D.J.: CASOS: a subspace method for anomaly detection in high dimensional astronomical databases. Stat. Anal. Data Min.: ASA Data Sci. J. **6**(1), 53–72 (2013). https://doi.org/10.1002/sam. 11167
13. Hunter, J.D.: Matplotlib: A 2D graphics environment. Comput. Sci. Eng. **9**, 90–95 (2007). https://doi.org/10.1109/MCSE.2007.55
14. Ishida, E.E.O., et al.: Optimizing spectroscopic follow-up strategies for supernova photometric classification with active learning. MNRAS **483**, 2–18 (2019). https:// doi.org/10.1093/mnras/sty3015
15. Ishida, E.E.O., et al.: Active Anomaly Detection for Time-Domain Discoveries. arXiv e-prints arXiv:1909.13260, September 2019
16. Jones, E., Oliphant, T., Peterson, P., et al.: SciPy: Open source scientific tools for Python (2001). http://www.scipy.org/
17. Lisakov, S.M., Dessart, L., Hillier, D.J., Waldman, R., Livne, E.: Progenitors of low-luminosity Type II-Plateau supernovae. MNRAS **473**, 3863–3881 (2018). https:// doi.org/10.1093/mnras/stx2521
18. Liu, F.T., Ting, K.M., Zhou, Z.H.: Isolation forest. In: 2008 Eighth IEEE International Conference on Data Mining, pp. 413–422. IEEE (2008)
19. Liu, F.T., Ting, K.M., Zhou, Z.H.: Isolation-based anomaly detection. ACM Trans. Knowl. Discov. Data **6**(1), 3:1–3:39 (2012).https://doi.org/10.1145/2133360. 2133363
20. LSST Science Collaboration, Abell, P.A., et al.: LSST Science Book, Version 2.0. ArXiv e-prints, December 2009
21. Malanchev, K.L., et al.: Use of machine learning for anomaly detection problem in large astronomical databases. In: Elizarov, A., Novikov, B., Stupnikov, S. (eds.) Data Analytics and Management in Data Intensive Domains: Selected Papers of the XXI International Conference on Data Analytics and Management in Data Intensive Domains, DAMDID/RCDL 2019. CEUR Workshop Proceedings 2523, pp. 205–216 (2019). http://ceur-ws.org/Vol-2523/paper20.pdf
22. McKinney, W.: Data structures for statistical computing in Python. In: van der Walt, S., Millman, J. (eds.) Proceedings of the 9th Python in Science Conference, pp. 51–56 (2010)
23. Moya, M.M., Hush, D.R.: Network constraints and multi-objective optimization for one-class classification. Neural Netw. **9**(3), 463–474 (1996). https://doi.org/10. 1016/0893-6080(95)00120-4. http://www.sciencedirect.com/science/article/pii/ 0893608095001204
24. Nakano, S., et al.: Supernova 2013am in M65 = PSN J11185695+1303494. Central Bureau Electronic Telegrams 3440, March 2013
25. Nun, I., Pichara, K., Protopapas, P., Kim, D.W.: Supervised detection of anomalous light curves in massive astronomical catalogs. APJ **793**, 23 (2014). https:// doi.org/10.1088/0004-637X/793/1/23
26. Pedregosa, F., et al.: Scikit-learn: machine learning in Python. J. Mach. Learn. Res. **12**, 2825–2830 (2011)
27. Pruzhinskaya, M.V., et al.: Anomaly detection in the open supernova catalog. Mon. Not. Roy. Astron. Soc. **489**(3), 3591–3608 (2019). https://doi.org/10.1093/mnras/ stz2362
28. Rasmussen, C.E., Williams, C.K.I.: Gaussian Processes for Machine Learning (Adaptive Computation and Machine Learning). The MIT Press, Cambridge (2005)

29. Rebbapragada, U., Protopapas, P., Brodley, C.E., Alcock, C.: Finding anomalous periodic time series. Mach. Learn. **74**(3), 281–313 (2009). https://doi.org/10.1007/s10994-008-5093-3

30. Sanders, N.E., et al.: PS1-12sk is a peculiar supernova from a He-rich progenitor system in a brightest cluster galaxy environment. APJ **769**, 39 (2013). https://doi.org/10.1088/0004-637X/769/1/39

31. Schölkopf, B., Platt, J.C., Shawe-Taylor, J.C., Smola, A.J., Williamson, R.C.: Estimating the support of a high-dimensional distribution. Neural Comput. **13**(7), 1443–1471 (2001). https://doi.org/10.1162/089976601750264965

32. Tarassenko, L., Hayton, P., Cerneaz, N., Brady, M.: Novelty detection for the identification of masses in mammograms (1995)

33. van der Walt, S., Colbert, S.C., Varoquaux, G.: The NumPy array: a structure for efficient numerical computation. Comput. Sci. Eng. **13**(2), 22–30 (2011). https://doi.org/10.1109/MCSE.2011.37

34. Wyrzykowski, L., et al.: Gaia16aye is a binary microlensing event and is crossing the caustic again. The Astronomer's Telegram 9507, September 2016

Modern Astronomical Surveys for Parameterization of Stars and Interstellar Medium

Oleg Malkov[1,2], Sergey Karpov[3,4,5], Dana Kovaleva[1], Jayant Murthy[6], Sergey Sichevsky[1], Nikolay Skvortsov[7(✉)], Sergey Stupnikov[7], Gang Zhao[2], and Aleksandr Zhukov[1,8,9]

[1] Institute of Astronomy, Moscow 119017, Russia
malkov@inasan.ru
[2] National Astronomical Observatories, Beijing 100012, China
[3] Institute of Physics, Czech Academy of Sciences, 182 21 Prague 8, Czech Republic
[4] Special Astrophysical Observatory, Nizhnij Arkhyz 36916, Russia
[5] Kazan Federal University, Kazan 420008, Russia
[6] Indian Institute of Astrophysics, Bengaluru 560034, India
[7] Institute of Informatics Problems, Federal Research Center "Computer Science and Control" of the Russian Academy of Sciences, Moscow 119333, Russia
nskv@mail.ru
[8] Sternberg Astronomical Institute, Moscow 119234, Russia
[9] Russian Technological University (MIREA), Moscow 119454, Russia
http://www.inasan.ru/~malkov

Abstract. Developing methods for analyzing and extracting information from modern sky surveys is a challenging task in astrophysical studies and is important for many investigations of galactic and extragalactic objects. We have designed a method for determination of stellar parameters and interstellar extinctions from multicolor photometry. This method was applied to objects drawn from modern large photometric surveys. In this work, we give a review of the surveys and discuss problems of cross-identification, paying particular attention to the information flags contained in the surveys. Also we have determined new statistical relations to estimate the stellar atmospheric parameters using MK spectral classification.

Keywords: Cross-matching · Sky surveys · Photometry · Interstellar extinction

1 Introduction

An outstanding problem of astrophysics is the study of the stellar physical properties. Because the stars are observed through interstellar dust, their light is dimmed and reddened, complicating their parameterization and classification. The parameters of a given star, as well as the interstellar reddening, may be

© Springer Nature Switzerland AG 2020
A. Elizarov et al. (Eds.): DAMDID/RCDL 2019, CCIS 1223, pp. 108–123, 2020.
https://doi.org/10.1007/978-3-030-51913-1_8

obtained from its spectrum but one must either use a large telescope or only observe bright objects in order to get spectral energy distributions with good resolution and sufficient accuracy. On the other hand, recently constructed large photometric surveys with new tools for cross-matching objects provide us with the possibility of getting multicolor photometric data for hundreds of millions of objects. From these, we may not only parameterize objects but also determine the 3-dimensional interstellar extinction in the Galaxy.

We have developed a method for the determination of stellar parameters and interstellar extinction values from multicolor photometry. The application of this method to a set of stars in a small area in the sky allows us to determine an increase of interstellar extinction with distance in that direction and, consequently, to construct a 3-d extinction map of the Milky Way Galaxy.

This work is an extension of [39]. Comparing it with the previous work in [39], the parameterization procedure (Sect. 3) is described in more detail, and the relations between spectral classes and atmospheric parameters (Table 1) are added; the description of the response curves of the photometric surveys is updated and improved (Fig. 1); Table 3 listing and describing the flags included in the surveys has been added; the list of coming photometric surveys (Sect. 6) is extended with the description of the VPHAS+ survey.

Published interstellar extinction maps are described in Sect. 2. Section 3 contains description of our procedure for parameterization of stars. In Sect. 4 we give a review of sky surveys, and present principles of their cross-matching. Our results are presented in Sect. 5. Our future plans are discussed in Sect. 6 with the conclusions in Sect. 7.

2   Interstellar Extinction Maps

Three-dimensional (3D) extinction models have been constructed using spectral and photometric stellar data, open cluster data, star counts, Galactic dust distribution models.

The standard approach to construct a 3D extinction model has been to parcel out the sky in angular cells, each defined by boundaries in Galactic coordinates (l, b). The visual extinction (A_V) in each cell may then be obtained as a function of distance (d): $A_V(l, b, d)$ from the stars in the cells. The angular size of the cells has varied from study to study, although each cell was generally chosen to be large enough to contain a statistically significant number of calibration stars at different distances.

Published 3D models, using spectral and photometric data, were based on 10^4-10^5 stars, or were constructed for a very limited area in the sky (see, e.g., [20, 27, 48], the earlier studies were reviewed in [37]). Modern large surveys contain photometric (3 to 5 bands) data for $10^7 - 10^9$ stars. However, to make those data (obtained at different wavelengths and with different observational techniques) useful for a 3D extinction model construction, one needs to run a correct cross-identification of objects between surveys. Such cross-identification was laborious and time consuming, but using Virtual Observatory (VO) data access and cross-correlation technologies, a search for counterparts in a subset of different catalogues can now be carried out in a few minutes. It is now feasible to obtain information on interstellar extinction from modern large photometric surveys.

To properly obtain astrophysical parameters from catalogued photometry one needs to study the possibility and sphere of application of the parameterization method. We indicate areas in the parameter space [effective temperature $\log T_{\text{eff}}$, gravity $\log g$, metallicity $[Fe/H]$, visual extinction A_V, total-to-selective extinction ratio R_V], where observational photometry precision, achieved in modern large multi-color surveys, allows us to obtain astrophysical parameters with acceptable accuracy [52].

3   Multicolor Photometry and Parameterization of Stars

3.1   Parameterization Procedure

We studied a problem of classification and parameterization of stars from multi-color photometry in detail (see, e.g., [53,54]). In particular, a problem of binary stars parameterization was studied in [40] and [38].

We have developed a method, which allows us to construct $A_V(l, b, d)$ relations from multicolor photometry. Varying (i) the spectral type of the star (SpT), (ii) its distance (d), and (iii) interstellar extinction value (A_V), we simulate the observational brightness, m, with the distance modulus equations

$$m_i = M_i(\text{SpT}) + 5 \log d - 5 + A_i(A_V) \tag{1}$$

for each of the i photometric bands included in the original surveys. Our goal is to minimize the function

$$\sigma = \sum_{i=1}^{N}((m_{obs,i} - m_i)/\delta m_{obs,i})^2, \tag{2}$$

and, based on the quality of the simulation process, choose the most appropriate SpT-d-A_V set. Here $m_{obs,i}$ and $\delta m_{obs,i}$ are catalogued magnitudes and their errors, respectively. A calibration relation $M_i(\text{SpT})$ and interstellar extinction law $A_i(A_V)$ should be available for each of the i photometric bands.

We have to remove from our simulation techniques all non-stellar objects, unresolved photometric binaries, variable stars and other contaminating objects, based on flags included in the original surveys. Low quality photometric data should also be eliminated from consideration. The list of flags included in the surveys is presented in Table 3 (see Sect. 4).

This method of simulation/parameterization, as described above, allows one to plot parameterized objects in the distance-extinction (d-A_V) plane, approximate them (by the cosecant law or more complicated function) and estimate interstellar extinction parameters in a given direction on the sky.

Note that for high galactic latitude areas ($|b| > 15°$ or so) the interstellar extinction is thought to be (roughly) uniformly distributed and to satisfy the so-called cosecant (barometric) law, suggested by Parenago in [42]. That function should be modified (complicated) for lower latitudes, as dust clouds concentrated in the Galactic plane, will have to be taken into account.

3.2 Relations Between Spectral Classes and Atmospheric Parameters

Absolute magnitude in i-th photometric band M_i in Eq. 1 depends on spectral type and is taken from calibration tables. To obtain absolute magnitudes for stars of different spectral classes and luminosity classes in the corresponding photometric systems M_i, we have used tables of absolute magnitudes in 2MASS, SDSS and GALEX systems [16, 25]. However, corresponding calibration tables, which provide stellar absolute magnitudes in a given photometric system for all spectral classes, can not be found in literature for other surveys (see the description of used surveys in Sect. 4). In the absence of such information, it is necessary to construct corresponding relations from theoretical spectral energy distributions (SED) and photometric system response curves. The best source for theoretical SEDs are libraries of synthetic spectra [6, 21, 30]. However, the SEDs are computed there for a given set of atmospheric parameters ($\log T_{\mathrm{eff}}$, $\log g$, $[M/H]$) rather than for spectral classes. Thus, for the solution of this problem it is necessary to design relations between spectral class and atmospheric parameters, for different luminosity classes.

We have constructed analytical statistical relations between MK spectral class and the atmospheric parameters (T_{eff}, $\log g$) for principal luminosity classes. To construct analytic (spectral class – atmospheric parameters) formula for main sequence stars we have used relations published in [14, 43, 57] and made a polynomial approximation. To construct analytic (spectral class – atmospheric parameters) formula for supergiant and giant stars we have used data from the empirical stellar spectral atlases ELODIE [46], Indo-US [59], MILES [15], and STELIB [29], and made a polynomial approximation. The results are presented in Table 1.

3.3 Modifications of the Procedure

Our procedure may be modified to use the astrometric and spectral information on the studied objects as input parameters. In particular, our procedure can be modified to determine stellar parameters and interstellar extinction values from not only multicolor photometry but also using additional information such as precise parallaxes and spectral classification, where available, thus reducing the number of unknowns in Eq. 1.

One notable improvement has come with the recent release of the Gaia DR2 (see Table 2) set of parallaxes, which allows us to use distance as an input (rather than as a free) parameter. It should significantly increase the accuracy of our results, especially when we can substitute the more precise parallaxes from Gaia DR3 for the DR2 data we currently use.

Table 1. Spectral class—effective temperature—surface gravity relations

		std. dev.	valid for	Eq.
LC=V				
$\log T_{\text{eff}}$	$= 4.80223 - 0.0465961S + 0.00157054S^2$	0.004	O3–O9	(1)
$\log T_{\text{eff}}$	$= 5.30408 - 0.111312S + 0.00284209S^2 -$ $-2.51285e^{-5}S^3$	0.011	B0–G7	
$\log T_{\text{eff}}$	$= 3.25745 + 0.0285452S - 0.000388153S^2$	0.008	G8–M9	
S	$= -77.4025 - 208.506T - 72.7616T^2$	0.36	$3.38 \leq \log T_{\text{eff}} < 3.75$	(2)
S	$= 13.0566 + 68.6827T + 404.486T^2 +$ $+751.0111T^3 + 497.913T^4$	0.75	$3.75 \leq \log T_{\text{eff}} < 4.10$	
S	$= 5.53554 - 34.2627T - 4.78570T^2 +$ $+191.168T^3 + 317.065T^4$	0.34	$4.10 \leq \log T_{\text{eff}} \leq 4.72$	
$\log g$	$= 4.23248 + 0.0194541S_1 +$ $+0.000552749S_1^2 - 4.30515e^{-5}S_1^3 -$ $-1.09920e^{-6}S_1^4 + 7.61843e^{-8}S_1^5 +$ $+8.20985e^{-10}S_1^6 - 3.27874e^{-11}S_1^7$	0.055	O3–M9.5	(3)
S	$= -0.117642 + 1.07059G + 192.069G^2 -$ $-183.386G^3 + 49.7143G^4$	4.02	$3.8 \leq \log g \leq 5.3$	(4)
LC=I				
$\log T_{\text{eff}}$	$= 5.37107 - 0.132197S + 0.00447197S^2 -$ $-7.12416e^{-5}S^3 + 4.17523e^{-7}S^4$	0.049	O7–M3	(5)
S	$= 5.87386 - 49.0805T - 135.952T^2 -$ $-119.090T^3 + 124.459T^4 + 108.708T^5$	3.14	$3.45 \leq \log T_{\text{eff}} < 4.60$	(6)
$\log g$	$= 5.26666 - 0.289286S + 0.00728099S^2 -$ $-6.33673e^{-5}S^3$	0.485	O7–M3	(7)
S	$= 5.26199 - 10.2492G + 2.79561G^2 +$ $+0.526251G^3$	9.74	$-0.2 \leq \log g \leq 3.8$	(8)
LC=III				
$\log T_{\text{eff}}$	$= 5.07073 - 0.0757056S + 0.00147089S^2 -$ $-1.03905e^{-5}S^3$	0.034	O5–M10	(9)
S	$= 8.49594 - 49.4053T - 191.524T^2 -$ $-335.488T^3 - 144.781T^4$	2.59	$3.45 \leq \log T_{\text{eff}} < 4.65$	(10)
$\log g$	$= 3.79253 - 0.0136260S +$ $+0.000562512S^2 - 1.68363e^{-5}S^3$	0.513	O5–M10	(11)
S	$= 33.3474 - 18.3022G -$ $-5.33024G^2 - 0.667234G^3$	7.03	$-0.5 \leq \log g \leq 4.7$	(12)

S is spectral class code: 3 for O3, ..., 10 for B0, ..., 60 for M0.
$S_1 = S - 35$
$T = \log T_{\text{eff}} - 4.6$
$G = \log g - 3.7$

Our procedure can also be modified for stars with spectral classification available from LAMOST [31], the largest source of spectral classification of objects in the northern sky. LAMOST Data Release 4 contains data on 7.6×10^6 objects and is available through VizieR database (V/153).

4 Sky Surveys and Cross-Matching

4.1 Sky Surveys Selection

The following sky surveys are selected for our study:

- The DENIS database [12];
- 2MASS All-Sky Catalog of Point Sources [9];
- The SDSS Photometric Catalogue, Release 12 [1];
- GALEX-DR5 (GR5) sources from AIS and MIS [4];
- UKIDSS-DR9 LAS, GCS and DXS Surveys [28];
- AllWISE Data Release [10];
- IPHAS DR2 Source Catalogue [3];
- The Pan-STARRS release 1 (PS1) Survey - DR1 [7];
- Gaia DR2 [2,19].

Some information on the surveys is given in Table 2, their photometric systems response curves are shown in Fig. 1 (the mid-IR AllWISE photometric bands are located in the 26,000–280,000 Å area and are not shown here).

Table 2. Large photometric surveys

Survey	Number of objects, 10^6	Sky coverage	Photometric bands	Limiting magnitude
DENIS	355	Southern hemisphere	Gunn-i, J, K_S	18.5, 16.5, 14.0
2MASS	471	All sky	J, H, K_S	15.8, 15.1, 14.3
SDSS 12	325	25%	u, g, r, i, z	g,r = 22.2
GALEX DR5 (AIS+MIS)	78	90%	FUV, NUV	~25
UKIDSS DR9 LAS	83	15%	Z, Y, J, H, K	K = 18.3
AllWISE	748	All sky	W1, W2, W3, W4	16.6, 16.0, 10.8, 6.7
IPHAS DR2	219	Northern Galactic plane	r, i, H_α	r = 21–22
Pan-STARRS PS1 - DR1	1919	All sky but southern cap	g, r, i, z, y	i~20
GAIA DR2	1693	All sky	G, BP, RP	G = 20

The selected surveys satisfy the following criteria:

- the number of objects exceeds 10×10^6;
- the survey covers a large area in the sky (the only exception is IPHAS, which covers a relatively small but important area in the sky);
- the photometric accuracy is better than about 0.05 mag;
- the depth of the survey exceeds V∼20 mag.

For every survey the following information should be available: absolute magnitude – spectral type (M_λ – SpT) calibration tables and $A_\lambda(A_V)$ relations for every photometric band λ. If these information is not available in literature, we construct it using response curves of photometric bands and spectral energy distribution (SED) for every spectral type, as well as the interstellar extinction law ([5,17,18]). Besides, relations between spectral type and atmospheric parameters (effective temperature $\log T_{\mathrm{eff}}$ and surface gravity $\log g$) for stars of different luminosity classes should be available.

To model observational photometry one needs to know spectral energy distribution, and a number of spectrophotometric atlases are designed to meet that requirement (e.g, [45,60]). We have made a comparative analysis of the most known semi-empirical and empirical spectral atlases. The results show that the standard error of synthesized stellar magnitudes calculated with SEDs from best spectral atlases reaches 0.02 mag. It has been also found that some modern spectral atlases are burdened with significant systematic errors [24].

A preliminary analysis of applicability of SDSS and 2MASS photometry for determining the properties of stars and interstellar extinction was made by in [51].

4.2   Cross-Matching of Surveys

The number of surveys available at any wavelength is large enough to construct detailed Spectral Energy Distributions (SEDs) for any kind of astrophysical object. However, different surveys/instruments have different positional accuracy and resolution. In addition, the depth of each survey is different and, depending on sources brightness and their SED, a given source might or might not be detected at a certain wavelength. All this makes the pairing of sources among catalogues not trivial, especially in crowded fields.

We have implemented an algorithm of fast positional matching of large astronomical catalogs in small (up to one degree) areas with filtering of false identification [34]. In particular, for each area and each pair we estimated the matching radius. As a result, we drew in a number 0.1-degree radius areas samples of point-like objects counterparts from the DENIS, 2MASS, SDSS, GALEX, and UKIDSS surveys, and performed a cross-identification within these surveys [23,33]. We have compiled the corresponding subcatalogues in the VOTable [41] format. The tool developed as a result of this work can be used to cross-identify objects in arbitrary sky areas for the further classification and determination of stellar parameters, including the measurement of the amount of interstellar extinction.

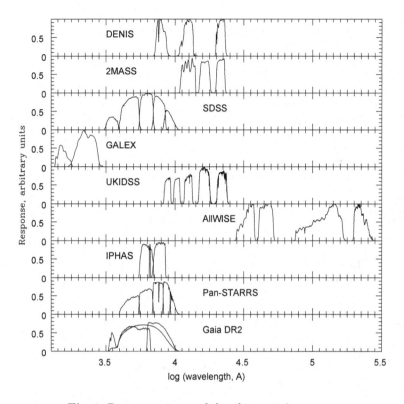

Fig. 1. Response curves of the photometric surveys

In some surveys (e.g., GALEX, SDSS, UKIDSS) more than one observation per object was made and, consequently, more than one entry per object is present in the catalogue. In such cases we use weighted average values for the photometry.

In the cross-identification process (and later for the parameterization) we use all positional information and all photometry available in surveys. To select objects for further study we also pay attention to various flags, presented in the surveys. The flags can indicate quality of observations and provide information on a nature of object (duplicity, variability, extended shape). As it was mentioned above, on this stage we do not use trigonometric parallax as an input parameter.

Response curves of photometric bands of the surveys are shown in Fig. 1. It can be seen that some bands in different surveys are the same or similar (e.g. K_S-band in DENIS and K_S-band in 2MASS). The comparison of brightness of objects in such pairs of bands provides us an additional filter to discard objects irrelevant for the parameterization: a large magnitude difference may indicate variability, a rare evolutionary stage, or non-stellar nature of the object. Too bright and too faint objects for this particular survey (i.e., overexposed and underexposed, respectively) can also be spotted and omitted at this stage.

Table 3. Flags in the surveys

Survey, flag	Meaning	Keep source/band	Eliminate source/band
2MASS			
Qflg[a]	Photometric quality	A, B, C, D	X, U, F, E
Rflg[a]	Source of photometry	1, 2 ,3	0, 4, 6, 9
Bflg[a]	Blend flag	1	0>1
Cflg[a]	Artifact contamination	0	non-zero values
Xflg	Extended source	0	1, 2
Aflg	Asteroid or comet	0	1
dup	Duplicate source	0 (and use = 1),1	0 (and use = 0),>1
use	Use source	see above	see above
GALEX[b]			
Fafl, Nafl	Artifact flag[c]	0, 1, >4	2, 4
nS/G, fS/G	Star/galaxy classifier	>0.5	≤0.5
SDSS			
class	Type of object	6	3
Q	Quality of observation	3	1, 2
UKIDSS[d]			
cl	Source class	−1, −2	−3, 0, 1
AllWISE			
ccf[e]	Contamination	0	d, p, h,o
ex	Extended source	0	>0
var[e]	Variability	n, 0, 1, 2, 3, 4, 5	6, 7, 8, 9
qph[e]	Photometric quality	A, B, C	U, X, Z
Pan-STARRS			
f ObjID[f]	Information flag	11–22, 25–28	0–10, 23, 24, 29, 30
Gaia			
Dup	Duplicated source	0	1
Var	Variability flag	CONSTANT, NOT AVAILABLE	VARIABLE
DENIS			
q Imag, q Jmag, q Kmag	Stellarity index	>75	≤75
Iflg, Jflg, Kflg	Image and source flag	1000, no flag	all others
IPHAS			
mergedClass	Image classification	−1, −2	1,0

[a] 2MASS Qflg, Rflg, Bflg, Cflg are 3-digit (or 3-letter) flags. Each digit (letter) refers to a corresponding photometric band: J, H, K_S.

[b] GALEX Fexf, Nexf (Extraction flags) and Fr, Nr (source FWHM flags) are stored, but not yet used.

[c] Note that if more than one GALEX artifact is deemed to be present, the flag value is the sum of all the artifacts affecting the source.

[d] UKIDSS p*, pG, pN probability flags are stored, but not yet used.

[e] AllWISE ccf, var, qph are 4-digit (or 4-letter) flags. Each digit (letter) refers to a corresponding photometric band: W1, W2, W3, W4.

[f] Pan-STARRS f_{objID} is the Information flag bitmask. The table gives the degrees of 2, i.e.,"11" means $2^{11} = 2048$. Note that if more than one flag is deemed to be present, the f_{objID} value is the sum of all the flags assigned to the source.

4.3   Selection of Sky Areas

To test our procedure, we have to select sky areas which are interesting from various astrophysical points of view and where our results can be compared with independent studies.

It is instructive and useful to apply the model to estimate interstellar extinction for several areas of the sky where individual estimates were made by [50], and used to calculate extinction for SNs in the Universe accelerating expansion study [44].

Among other interesting objects, RR Lyr-type variable stars (variables) were selected for the study. RR Lyr-type pulsating variables satisfy a period-luminosity relation (PLR) that simplifies estimation of their distances (and, consequently, distances to stellar systems they reside). However, PLR is not yet well calibrated, and our study of dust distribution in the RR Lyr-type variables directions is intended to improve the situation. Several hundreds of RR Lyr-type variables with available spectral classification were selected for our study from the General Catalogue of Variable Stars [49].

Another interesting direction in the sky to study is the solar apex, i.e., the direction that the Sun travels with respect to the mean motion of material in the Solar neighborhood. The solar apex is in the constellation of Hercules, the approximate galactic coordinates are l = 56°.24, b = 22°.54. There is a practical interest in the study of dust distribution in the Galaxy in that direction. The movement of the Solar system through the clots of interstellar gas could lead to the direct invasion of a dense mixture of gas and dust into the Solar system. That has such potential consequences as global glaciation and reducing the size of the heliosphere (up to the Earth's orbit) which protects us from cosmic rays.

5   Results and Discussion

In our pilot study [35] we applied this method to construct $A_V(l, b, d)$ relations for selected areas at high galactic latitudes. We have cross-matched objects in 2MASS, SDSS, GALEX and UKIDSS surveys in selected areas in the sky, using Virtual Observatory facilities. As a result of the cross-matching, we find multi-wavelength ($i = 9$ to 13 bands) photometric data for each object.

We have compared our results with LAMOST [31] data and extinction values to distant SNs (based on IRAS and DIRBE microwave data), available in the literature. The comparison exhibits a good agreement (see [35] for details). A comparison of our results with recently released Gaia DR2 data also demonstrates a good agreement for stars as faint as $19^m.6$ g_{SDSS}, and shows that our method allows us to determine spectral type, distance and interstellar extinction of objects out to 4.5 kpc [36]. It indicates that the proposed algorithm (after some modifications, required for low galactic latitudes) can be used for construction of a 3D map of interstellar extinction in the Milky Way Galaxy.

6 Future Plans

6.1 Coming Photometric Surveys

Our experience is thought to be a practical guide to issues that will be particular important as soon as the new surveys will become available. In particular, the following surveys can be mentioned here.

LSST. Legacy Survey of Space and Time (LSST) is the most ambitious survey currently planned in the optical [22]. LSST will be a large, wide-field ground-based system designed to obtain repeated images covering the sky visible from northern Chile. The telescope will have an 8.4 m (6.5 m effective) primary mirror, a 9.6 deg^2 field of view, a 3.2-gigapixel camera, and six filters (ugrizy) covering the wavelength range 320–1050 nm. The project is in the construction phase and will begin regular survey operations by 2022. A 18,000 deg^2 region will be uniformly observed during the anticipated 10 yr of operations and will yield a co-added map to r $\sim$ 27.5. These data will result in databases including about 32 trillion observations of 20 billion galaxies and a similar number of stars, and they will serve the majority of the primary science programs.

SAGE. Stellar Abundance and Galactic Evolution (SAGE) project aims to study the stellar atmospheric parameters of $\sim$0.5 $\times 10^9$ stars in the $\sim$12.000 deg^2 of the northern sky, with declination $\delta > -5°$, excluding the bright Galactic disk ($|b| < 10°$) and the sky area of $12 < R.A. < 18$ hr [62]. The survey uses a self-designed SAGE photometric system, which is composed of eight photometric bands: Stromgren-u, SAGE-v, SDSS g,r,i, $H_\alpha$wide, $H_\alpha$narrow, and DDO-51.

UVIT. The UVIT instrument on-board the Indian space observatory ASTROSAT consists of two 38-cm telescopes—one for the FUV and the other for the NUV and visible bands. It has a circular field of view $\approx 28'$ in diameter. It collects data in three channels simultaneously, in FUV, NUV and Visible bands corresponding to $\lambda = 1300$–1800 A, 2000–3000 A and 3200–5500 A, respectively. Full details of the instrument and calibration results can be found in [58]. UVIT does not provide data for *large* number of objects, however, its data will be used as the UV spectral range is very important for the study of the interstellar extinction.

VPHAS+. The VST Photometric Hα Survey of the Southern Galactic Plane and Bulge (VPHAS+) is surveying the southern Milky Way in u, g, r, i and Hα at $\sim$1" angular resolution [13]. Its footprint spans the Galactic latitude range $-5° < b < +5°$ at all longitudes south of the celestial equator. Extensions around the Galactic Centre to Galactic latitudes $\pm 10°$ bring in much of the Galactic Bulge. This ESO public survey, begun on 28th December 2011, reaches down to $\sim$20th magnitude (10σ) and will provide single-epoch digital optical photometry for $\sim$300 million stars.

Another aspect which we can tackle is how the accuracy of the results depend on missing data (in fact the larger the number of the surveys cross-matched, the larger should be the fraction of missing data). According to our preliminary results [35], the presence or absence of 2MASS data in the set (subject to the availability of SDSS, GALEX and UKIDSS data) does not significantly change the result, but this issue needs a further study.

6.2   Use of Spectral Surveys in Parameterization

Ongoing (LAMOST [32], APOGEE (all-sky, ~450,000 objects) [47], SEGUE (northern sky, ~350,000 objects) [61], RAVE (southern sky, ~460,000 objects) [26] and upcoming (4MOST [11], MOONS [8], WEAVE [56]) spectroscopic surveys can serve as an exceptional sources not only of stellar parameter values, but also of the nature of interstellar dust and its distribution in the Milky Way. Atmospheric parameters (T_{eff}, $\log g$) and/or spectral classifications—obtained from spectroscopy combined with observational photometry—allow us to determine distances and interstellar extinctions for stars with high accuracy and thereby to construct a 3D map of interstellar extinction.

7   Conclusion

The parameterization of stars is a well known problem and used for various purposes in astronomy (e.g., while solving the problem of searching for well defined stars to be used for secondary photometric standards [55]). We have shown that multicolor photometric data from large modern surveys can be used for parameterization of stars. A comparison of our results with independent data shows a good agreement. We prove that with sufficiently good quality photometry, one may compute a 3D extinction map by comparing catalogued multicolor photometry with photometry derived from the secondary estimators such as the distance modulus and the interstellar extinction law with suitable calibration tables for absolute magnitudes with reasonable spectral types, extinctions and distances.

With the advent of large, existing and coming, photometric surveys and the evolution of computing power and data analysis techniques (in particular, Virtual Observatory tools for cross-matching), interstellar extinction can now be computed for hundreds of millions of stars in a reasonable amount of time, and a 3D interstellar extinction map can be constructed.

Acknowledgement. We thank our reviewers whose comments greatly helped us to improve the paper. OM thanks the CAS President's International Fellowship Initiative (PIFI). The work was partly supported by the Russian Foundation for Basic Researches (projects 17-52-45076, 18-07-01434, 19-07-01198) and by DST grant INT/RUS/RFBR/P-265 to JM. S.K. was supported by European Structural and Investment Fund and the Czech Ministry of Education, Youth and Sports (Project CoGraDS – CZ.02.1.01/0.0/0.0/15 003/0000437). The work is partially performed according to the Russian Government Program of Competitive Growth of Kazan Federal University. This research has made use of NASA's Astrophysics Data System, and use of the VizieR catalogue access tool, CDS, Strasbourg, France.

References

1. Alam, A., et al.: The eleventh and twelfth data releases of the sloan digital sky survey: final data from SDSS-III. Astrophys. J. Suppl. Ser. **219**, 12 (2015). https://doi.org/10.1088/0067-0049/219/1/12
2. Bailer-Jones, C.A.L., Rybizki, J., Fouesneau, M., Mantelet, G., Andrae, R.: Estimating distance from parallaxes. IV. distances to 1.33 billion stars in gaia data release 2. Astron. J. **156**, 58 (2018). https://doi.org/10.3847/1538-3881/aacb21
3. Barentsen, G., et al.: VizieR online data catalog: IPHAS DR2 source catalogue (Barentsen+, 2014). VizieR Online Data Catalog, **2321** (2014)
4. Bianchi, L., Shiao, B., Thilker, D.: Revised catalog of GALEX ultraviolet sources. I. The all-sky survey: GUVcat_AIS. Astrophys. J. Suppl. Ser. **230**, 24 (2017). https://doi.org/10.3847/1538-4365/aa7053
5. Cardelli, J.A., Clayton, G.C., Mathis, J.S.: The relationship between infrared, optical, and ultraviolet extinction. Astrophys. J. **345**, 245–256 (1989). https://doi.org/10.1086/167900
6. Castelli, F., Kurucz, R.L.: New grids of ATLAS9 model atmospheres. In: Piskunov, N., Weiss, W.W., Gray, D.F. (eds.) Modelling of Stellar Atmospheres, IAU Symposium, vol. 210, p. A20 (2003)
7. Chambers, K.C., et al.: The Pan-STARRS1 surveys. arXiv e-prints (2016)
8. Cirasuolo, M.: MOONS consortium: MOONS: a new powerful multi-object spectrograph for the VLT. In: Skillen, I., Balcells, M., Trager, S. (eds.) Multi-Object Spectroscopy in the Next Decade: Big Questions, Large Surveys, and Wide Fields, Astronomical Society of the Pacific Conference Series, vol. 507, p. 109 (2016)
9. Cutri, R.M., et al.: VizieR online data catalog: 2MASS all-sky catalog of point sources (Cutri+ 2003). VizieR Online Data Catalog, **2246** (2003)
10. Cutri, R.M., et al.: VizieR online data catalog: AllWISE data release (Cutri+ 2013). VizieR Online Data Catalog, **2328** (2014)
11. de Jong, R.S., et al.: 4MOST: project overview and information for the first call for proposals. Messenger **175**, 3–11 (2019). https://doi.org/10.18727/0722-6691/5117
12. DENIS Consortium: VizieR online data catalog: the DENIS database (DENIS Consortium, 2005). VizieR Online Data Catalog, **1** (2005)
13. Drew, J.E., et al.: The VST photometric Hα survey of the southern galactic plane and bulge (VPHAS+). Mon. Not. R. Astron. Soc. **440**(3), 2036–2058 (2014). https://doi.org/10.1093/mnras/stu394
14. Eker, Z., et al.: Interrelated main-sequence mass-luminosity, mass-radius, and mass-effective temperature relations. Mon. Not. R. Astron. Soc. **479**, 5491–5511 (2018). https://doi.org/10.1093/mnras/sty1834
15. Falcón-Barroso, J., et al.: An updated MILES stellar library and stellar population models. Astron. Astrophys. **532**, A95 (2011). https://doi.org/10.1051/0004-6361/201116842
16. Findeisen, K., Hillenbrand, L., Soderblom, D.: Stellar activity in the broadband ultraviolet. Astron. J. **142**, 23 (2011). https://doi.org/10.1088/0004-6256/142/1/23
17. Fitzpatrick, E.L., Massa, D.: An analysis of the shapes of interstellar extinction curves. V. The IR-through-UV curve morphology. Astrophys. J. **663**, 320–341 (2007). https://doi.org/10.1086/518158
18. Fluks, M.A., Plez, B., The, P.S., de Winter, D., Westerlund, B.E., Steenman, H.C.: On the spectra and photometry of M-giant stars. Astron. Astrophys. Suppl. Ser. **105**, 311–336 (1994)

19. Gaia Collaboration, Brown, A.G.A., et al.: Gaia data release 2. Summary of the contents and survey properties. Astron. Astrophys. **616**, A1 (2018). https://doi.org/10.1051/0004-6361/201833051

20. Green, G.M., et al.: A three-dimensional map of milky way dust. Astrophys. J. **810**, 25 (2015). https://doi.org/10.1088/0004-637X/810/1/25

21. Gustafsson, B., Edvardsson, B., Eriksson, K., Jørgensen, U.G., Nordlund, Å., Plez, B.: A grid of MARCS model atmospheres for late-type stars. I. Methods and general properties. Astron. Astrophys. **486**, 951–970 (2008). https://doi.org/10.1051/0004-6361:200809724

22. Ivezić, Ž., et al.: LSST: from science drivers to reference design and anticipated data products. Astrophys. J. **873**, 111 (2019). https://doi.org/10.3847/1538-4357/ab042c

23. Karpov, S.V., Malkov, O.Y., Mironov, A.V.: Cross-identification of large surveys for finding interstellar extinction. Astrophys. Bull. **67**, 82–89 (2012). https://doi.org/10.1134/S1990341312010087

24. Kilpio, E.Y., Malkov, O.Y., Mironov, A.V.: Comparative analysis of modern empirical spectro-photometric atlases with multicolor photometric catalogues. In: Prugniel, P., Singh, H.P. (eds.) Astronomical Society of India Conference Series, vol. 6, p. 31 (2012)

25. Kraus, A.L., Hillenbrand, L.A.: The stellar populations of praesepe and coma berenices. Astron. J. **134**, 2340–2352 (2007). https://doi.org/10.1086/522831

26. Kunder, A., et al.: The radial velocity experiment (RAVE): fifth data release. Astron. J. **153**, 75 (2017). https://doi.org/10.3847/1538-3881/153/2/75

27. Lallement, R., et al.: Three-dimensional maps of interstellar dust in the local arm: using Gaia, 2MASS, and APOGEE-DR14. Astron. Astrophys. **616**, A132 (2018). https://doi.org/10.1051/0004-6361/201832832

28. Lawrence, A., et al.: The UKIRT infrared deep sky survey (UKIDSS). Mon. Not. R. Astron. Soc. **379**, 1599–1617 (2007). https://doi.org/10.1111/j.1365-2966.2007.12040.x

29. Le Borgne, J.F., et al.: STELIB: a library of stellar spectra at R~3000. Astron. Astrophys. **402**, 433–442 (2003). https://doi.org/10.1051/0004-6361:20030243

30. Lejeune, T., Cuisinier, F., Buser, R.: Standard stellar library for evolutionary synthesis. I. Calibration of theoretical spectra. Astron. Astrophys. Suppl. Ser. **125**, 229–246 (1997). https://doi.org/10.1051/aas:1997373

31. Luo, A.L., et al.: The first data release (DR1) of the LAMOST regular survey. Res. Astron. Astrophys. **15**, 1095 (2015). https://doi.org/10.1088/1674-4527/15/8/002

32. Luo, A.L., Zhao, Y.H., Zhao, G., et al.: VizieR online data catalog: LAMOST DR5 catalogs (Luo+, 2019). VizieR Online Data Catalog V/164 (2019)

33. Malkov, O., et al.: Cross catalogue matching with virtual observatory and parametrization of stars. Baltic Astron. **21**, 319–330 (2012). https://doi.org/10.1515/astro-2017-0390

34. Malkov, O., Karpov, S.: Cross-matching large photometric catalogs for parameterization of single and binary stars. In: Evans, I.N., Accomazzi, A., Mink, D.J., Rots, A.H. (eds.) Astronomical Data Analysis Software and Systems XX. Astronomical Society of the Pacific Conference Series, vol. 442, p. 583 (2011)

35. Malkov, O., et al.: Interstellar extinction from photometric surveys: application to four high-latitude areas. Open Astron. **27**, 62–69 (2018). https://doi.org/10.1515/astro-2018-0002

36. Malkov, O., et al.: Verification of photometric parallaxes with gaia DR2 data. Galaxies **7**, 7 (2018). https://doi.org/10.3390/galaxies7010007

37. Malkov, O., Kilpio, E.: A synthetic map of the galactic interstellar extinction. Astrophys. Space Sci. **280**, 115–118 (2002). https://doi.org/10.1023/A: 1015526811574

38. Malkov, O., Mironov, A., Sichevskij, S.: Single-binary star separation by ultraviolet color index diagrams. Astrophys. Space Sci. **335**, 105–111 (2011). https://doi.org/ 10.1007/s10509-011-0613-1

39. Malkov, O.Y., et al.: Cross-matching of objects in large sky surveys. In: CEUR Workshop Proceedings, vol. 2523, pp. 217–228 (2019)

40. Malkov, O.Y., Sichevskij, S.G., Kovaleva, D.A.: Parametrization of single and binary stars. Mon. Not. R. Astron. Soc. **401**, 695–704 (2010). https://doi.org/ 10.1111/j.1365-2966.2009.15696.x

41. Ochsenbein, F., et al.: VOTable: tabular data for the virtual observatory. In: Quinn, P.J., Górski, K.M. (eds.) Toward an International Virtual Observatory. ESO, pp. 118–123. Springer, Heidelberg (2004). https://doi.org/10.1007/10857598_18

42. Parenago, P.P.: On interstellar extinction of light. Astron. Zh. **13**, 3 (1940)

43. Pecaut, M.J., Mamajek, E.E.: Intrinsic colors, temperatures, and bolometric corrections of pre-main-sequence stars. Astrophys. J. Suppl. Ser. **208**, 9 (2013). https://doi.org/10.1088/0067-0049/208/1/9

44. Perlmutter, S., et al.: Measurements of Ω and Λ from 42 high-redshift supernovae. Astrophys. J. **517**, 565–586 (1999). https://doi.org/10.1086/307221

45. Pickles, A.J.: A stellar spectral flux library: 1150–25000 Å. Publ. Astron. Soc. Pac. **110**, 863–878 (1998). https://doi.org/10.1086/316197

46. Prugniel, P., Soubiran, C., Koleva, M., Le Borgne, D.: New release of the ELODIE library: Version 3.1. arXiv Astrophysics e-prints (2007)

47. Reis, I., Poznanski, D., Baron, D., Zasowski, G., Shahaf, S.: Detecting outliers and learning complex structures with large spectroscopic surveys - a case study with APOGEE stars. Mon. Not. R. Astron. Soc. **476**, 2117–2136 (2018). https://doi. org/10.1093/mnras/sty348

48. Sale, S.E., et al.: A 3D extinction map of the northern Galactic plane based on IPHAS photometry. Mon. Not. R. Astron. Soc. **443**, 2907–2922 (2014). https:// doi.org/10.1093/mnras/stu1090

49. Samus', N.N., Kazarovets, E.V., Durlevich, O.V., Kireeva, N.N., Pastukhova, E.N.: General catalogue of variable stars: version GCVS 5.1. Astron. Rep. **61**, 80–88 (2017). https://doi.org/10.1134/S1063772917010085

50. Schlegel, D.J., Finkbeiner, D.P., Davis, M.: Maps of dust infrared emission for use in estimation of reddening and cosmic microwave background radiation foregrounds. Astrophys. J. **500**, 525–553 (1998). https://doi.org/10.1086/305772

51. Sichevskij, S.G.: Applicability of broad-band photometry for determining the properties of stars and interstellar extinction. Astrophys. Bull. **73**, 98–107 (2018). https://doi.org/10.1134/S199034131801008X

52. Sichevskij, S.G., Mironov, A.V., Malkov, O.Y.: Accuracy of stellar parameters determined from multicolor photometry. Astrophys. Bull. **69**, 160–168 (2014). https://doi.org/10.1134/S1990341314020035

53. Sichevskiy, S.G., Mironov, A.V., Malkov, O.Y.: Classification of stars with WBVR photometry. Astron. Nachr. **334**, 832 (2013). https://doi.org/10.1002/ asna.201311932

54. Sichevsky, S., Malkov, O.: Estimating stellar parameters and interstellar extinction from evolutionary tracks. Baltic Astron. **25**, 67–74 (2016). https://doi.org/ 10.1515/astro-2017-0112

55. Skvortsov, N.A., et al.: Conceptual approach to astronomical problems. Astrophys. Bull. **71**(1), 114–124 (2016). https://doi.org/10.1134/S1990341316010120

56. Smith, D.J.B., et al.: The WEAVE-LOFAR survey. In: Reylé, C., et al. (eds.) SF2A-2016: Proceedings of the Annual Meeting of the French Society of Astronomy and Astrophysics, pp. 271–280 (2016)

57. Straižys, V.: Multicolor stellar photometry (1992)

58. Tandon, S.N., et al.: In-orbit calibrations of the ultraviolet imaging telescope. Astron. J. **154**, 128 (2017). https://doi.org/10.3847/1538-3881/aa8451

59. Valdes, F., Gupta, R., Rose, J.A., Singh, H.P., Bell, D.J.: The Indo-US library of Coudé feed stellar spectra. Astrophys. J. Suppl. Ser. **152**, 251–259 (2004). https://doi.org/10.1086/386343

60. Wu, Y., Singh, H.P., Prugniel, P., Gupta, R., Koleva, M.: Coudé-feed stellar spectral library - atmospheric parameters. Astron. Astrophys. **525**, A71 (2011). https://doi.org/10.1051/0004-6361/201015014

61. Yanny, B., et al.: SEGUE: a spectroscopic survey of 240,000 stars with g = 14–20. Astron. J. **137**, 4377–4399 (2009). https://doi.org/10.1088/0004-6256/137/5/4377

62. Zheng, J., et al.: The SAGE photometric survey: technical description. Res. Astron. Astrophys. **18**, 147 (2018). https://doi.org/10.1088/1674-4527/18/12/147

Searching for Optical Counterparts of LIGO/Virgo Events in O2 Run

Elena Mazaeva[1]([✉]), Alexei Pozanenko[1], Alina Volnova[1], Pavel Minaev[1][ID],
Sergey Belkin[1], Raguli Inasaridze[2,3], Evgeny Klunko[4], Anatoly Kusakin[5],
Inna Reva[5], Vasilij Rumyantsev[6], Artem Novichonok[7,8], Alexander Moskvitin[9],
Gurgen Paronyan[10], Sergey Schmalz[7], and Namkhai Tungalag[11]

[1] Space Research Institute (IKI), 84/32 Profsoyuznaya Str, 117997 Moscow, Russia
`elena.mazaeva@phystech.edu`
[2] Kharadze Abastumani Astrophysical Observatory, Ilia State University,
0162 Tbilisi, Georgia
[3] Samtskhe-Javakheti State University, 113 Rustaveli Str., 0080 Akhaltsikhe, Georgia
[4] Institute of Solar Terrestrial Physics, 664033 Irkutsk, Russia
[5] Fesenkov Astrophysical Institute, 050020 Almaty, Kazakhstan
[6] Crimean Astrophysical Observatory, 298409 Nauchny, Crimea, Russia
[7] Keldysh Institute of Applied Mathematics of Russian Academy of Sciences,
4 Miusskaya Str, 125047 Moscow, Russia
[8] Petrozavodsk State University, 33 Lenina Str, 185910 Petrozavodsk, Russia
[9] Special Astrophysical Observatory of Russian Academy of Sciences,
369167 Nizhniy Arkhyz, Russia
[10] Byurakan Astrophysical Observatory, 0213 Byurakan, Aragatzotn Province,
Republic of Armenia
[11] Institute of Astronomy and Geophysics of Mongolian Academy of Sciences,
13343 Ulaanbaatar, Mongolia

Abstract. The problem of searching for optical counterparts of gravitational wave LIGO/Virgo events is discussing. Multi-messenger astronomy boosts the use a huge amount of astronomical data obtained by virtually all observatories around the world. We are discussing different tactics used for observations, problem of search for transients in the extremely large localization error-box of LIGO/Virgo events, and lessons obtained during second observational run of LIGO/Virgo in 2017. In particular we present our experience and results of follow up observations of LIGO/Virgo optical counterpart candidates by Space Research Institute (IKI) Gamma-Ray Bursts Follow up Network (IKI-GRB-FuN).

Keywords: Multi-messenger astronomy · Gravitational wave events ·
LIGO/Virgo · Gamma-ray bursts · Afterglow · Kilonova · Photometry

1 Introduction

The problem of search and observations of new transient objects is one of the main problems in modern astrophysics. It requires wide-field coverage and following comparison with available catalogues. Dedicated surveys and experiments

© Springer Nature Switzerland AG 2020
A. Elizarov et al. (Eds.): DAMDID/RCDL 2019, CCIS 1223, pp. 124–143, 2020.
https://doi.org/10.1007/978-3-030-51913-1_9

produce huge amounts of data daily in every domain of electromagnetic spectrum: in high energy range [34], optics [19,37], and radio [16], as well as in neutrino [22] and gravitational waves [27,54]. The surveys of new generation, like Large Synoptic Survey Telescope [32], will produce data of unprecedented volume and complexity. Reduction and analysis of these enormous data sets is already out of human's capacity and is similar to a search of a needle in a haystack. This problem is also connected with the search of transients related to the gravitational waves detections in very large localization areas, provided by LIGO and Virgo observations during theirs third scientific observational run in 2019–2020.

The Laser Interferometer Gravitational-Wave Observatory (LIGO) is designed to open the field of gravitational-wave astrophysics through the direct detection of gravitational waves predicted by Einstein's General Theory of Relativity [28]. LIGO's multikilometer-scale gravitational wave detectors use laser interferometry to measure the minute ripples in space-time caused by passing gravitational waves from cataclysmic cosmic events such as merging neutron stars (NSs) or black holes (BHs), or by supernovae. LIGO consists of two widely separated interferometers within the United States – one in Hanford, Washington and the other in Livingston, Louisiana – operated in unison to detect gravitational waves.

The first success of LIGO observations came in 2015 with the first direct observations of gravitational waves from the binary black hole merging GW150914 [9]. In 2017, when the sensitivity of LIGO detectors increased, and Virgo detector in Italy started its first observational cycle [10], the merging of binary neutron star was detected for the first time [1].

In the context of gravitational waves detection, one of the important problems for astrophysics are the search, identification, and observations of the possible electromagnetic (EM) counterpart of the event. Theory predicts no EM radiation from the binary BH coalescence since there is no matter outer the source that can produce it. In practice, there may be some radiation caused by accretion of a circumstellar matter on the resulting black hole, but its predicted flux is extremely low (e.g. [7]). Quite different situation is the binary NS merging (BNS). In this case, the merging objects consist of an ordinary matter that may produce high-energy EM radiation process (short gamma-ray burst) after the merging BNS an afterglow of wide energy range, and most interesting BNS counterpart which is called 'kilonova'.

The association of BNS merging with short gamma-ray bursts and kilonovae was first predicted theoretically [26], and then was confirmed observationally with the detection of GW170817/GRB170817A/AT2017gfo [2]. Besides the fact that GW170817 was the first case of the registration of gravitational waves from a BNS merging, it was also the first detection of gravitational waves and EM radiation from the same source [42].

A signal from the binary system merging is modeled numerically based on the Einstein's General Theory of Relativity and represents a package of oscillations increasing in amplitude with a decreasing period. The processing algorithm of

LIGO and Virgo detectors searches for modeled templates in the received data using different analyses. Localization on the sky is performed by triangulation method, measuring the time lag between the detection time for spatially distributed detectors, which determines the sky area of the most probable localization of the source. The time of the signal registration is measured with a high accuracy; however, the localization area may be very large, tens to hundreds of square degrees (see Table 1). GW170817 [49] has a localization region of ~30 square degrees, and there were reported ~190 galaxies in the volume limited by the sky area and distance estimates [13]. The kilonova AT2017gfo was discovered independently by 6 survey projects and was observed during several dozens of days in wide energy range from X-rays to radio [53]. The co-authors of the paper used the mosaic method to search for optical counterpart, observed of kilonova and proposed the model of prompt emission of GW170817/GRB 170817A [35].

In this paper, we discuss the problem of the search of a new transient optical source in large areas provided by detections of gravitational wave sources. We describe two basic methods of the search: mosaic observations of localization area and pre-determined target observations, i.e. search for transients in galaxies inside the detection volume. We also provide several examples of such semi-manual searches using available ground-based optical telescopes performed during the LIGO/Virgo observational run O2. Multi-messenger Astronomy is becoming a commonplace.

Table 1. Selected source parameters of the eleven confirmed GW detections of observational run O2 [3].

Event	Type	d_L/Mpc	$\Delta\Omega/deg^2$
GW150914	BBH	$430\pm^{150}_{170}$	180
GW151012	BBH	$1060\pm^{540}_{480}$	1555
GW151226	BBH	$440\pm^{180}_{190}$	1033
GW170104	BBH	$960\pm^{430}_{410}$	924
GW170608	BBH	$320\pm^{120}_{110}$	396
GW170729	BBH	$2750\pm^{1350}_{1320}$	1033
GW170809	BBH	$990\pm^{320}_{380}$	340
GW170814	BBH	$580\pm^{160}_{210}$	87
GW170817	BNS	$40\pm^{10}_{10}$	16
GW170818	BBH	$1020\pm^{430}_{360}$	39
GW170823	BBH	$1850\pm^{840}_{840}$	1651

BBH – binary black holes,
BNS – binary neutron stars,
d_L – luminosity distance,
$\Delta\Omega$ – error box of sky localization.

The paper continues the work [33]. We analyze several additional events (the results section). For all of these events, results of optical observations are presented: PS17dp, PS17fl, PS17fn, PS17gl of LIGO/Virgo event GW170104, PS17yt of LIGO/Virgo candidate GW170120_G270580, PS17lk, PS17qk, PS17nv, PS17pv, PS17rc, MASTER OT 90737.22+611200.5 of LIGO/Virgo candidate GW170120_G270580, PS17bek of LIGO/Virgo candidate GW170217_G268556, PS17bub, PS17bue of LIGO/Virgo candidate GW170227_G275697, GWFUNC-17ure of LIGO/Virgo event GW170823 and MASTER OT 33744.97+723159.0 of LIGO/Virgo candidate GW170825_G299232.

2   The Optical Transient Search Procedure

After receiving the alert signal from LIGO/Virgo, observations are carried out on ground-based optical telescopes to search for counterpart.

One can observe the whole area of localization with wide-field telescopes. This observation tactic is suitable if the localization area is not very large (up to about one hundred square degrees), or it is possible to observe with a large number of telescopes.

Since we know not only the localization region in the celestial sphere of the gravitational-wave event, but also the distance to the source, we can perform observations of galaxies within the gravitational wave localization volume, which is the second tactic. For this purpose, there is a value-added full-sky catalogue of galaxies, named as Galaxy List for the Advanced Detector Era, or GLADE [14]. GLADE was constructed by cross-matching and combining data from five separate (but not independent) astronomical catalogues: GWGC, 2MPZ, 2MASS XSC, HyperLEDA, and SDSS-DR12Q. But GLADE is complete up only to $D_L = 37(+3/-4)$ Mpc in terms of the cumulative B-band luminosity of galaxies within luminosity distance D_L, and contains all of the brightest galaxies giving half of the total B-band luminosity up to $D_L = 91$ Mpc, while the distance to the registered LIGO/Virgo source can reach several thousand Mpc (see Table 1).

All discussed tactics suggest a search for a transient on the obtained optical images. To search for optical transients we use the procedure of comparison of the generated catalog of sources selected from the image with all-sky catalogs. A block diagram of the method is presented in Fig. 1.

Block 1 – detecting and classifying of sources from astronomical images - the creation of an object catalog. In this case, we used SExtractor – software for source extraction [6]. Next, the astrometry procedure is necessary (e.g., using Apex [15] or using the requests library of astrometry.net [25]). To avoid incorrect cross-identification, we reject objects at the border of a frame (either at a distance of <4 FWHM from the border or using the value of the SExtractor flags). The current sensitivity of the LIGO/Virgo detectors allows to detect BNS and NS+BH mergers in relatively nearby galaxies (up to 500 Mpc). Such galaxies are registered typically as extended objects on an optical images, while the mergers alone represent point sources. Therefore, the task of extracting the point sources on the background of extended objects is very actual.

Block 2 – constructing the catalog of the comparison stars. For comparison of objects we use photometrical catalogs (e.g. SDSS, Pan-STARRS, APASS,

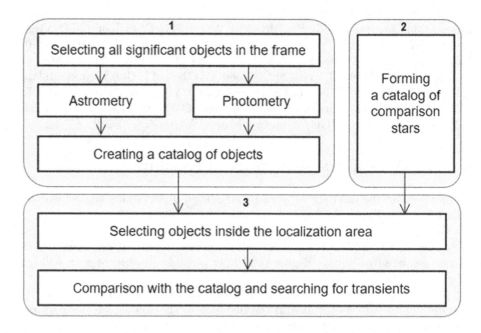

Fig. 1. A block diagram of the match algorithm.

2MASS, it depends on the optical filter of the original image, the image upper limit and the region of the celestial sphere).

Block 3 – searching for transient sources. Comparison of objects is performed using equatorial coordinates and magnitudes simultaneously within the measurement errors.

3   Results

The optical data was obtained by IKI GRB Follow-up Network, which is based at the Space Research Institute and provided follow-up observations of gravitational wave events in the optical range during Second Observing Run of LIGO/Virgo detectors. IKI GRB Follow-up Network collaborates with Crimean Astrophysical Observatory (CrAO) with ZTSh-2.6m telescope, Sayan Solar Observatory (Mondy) with AZT-33IK telescope, Tian Shan Astrophysical Observatory (TShAO) with Zeiss-1000 telescope, Abastumani Astrophysical Observatory (AbAO) with AS-32 telescope, Maidanak High-altitude Astronomical Observatory with AZT-22 telescope, ISON-Khureltogoot with VT-78a and ORI-40 telescopes, ISON-Castelgrande Observatory with ORI-22 telescope, Simeiz/Koshka observatory of INASAN with Zeiss-1000 telescope and Byurakan Astrophysical Observatory (BAO) with ZTA-2.6m telescope. Also some optical data were obtained with Zeiss-1000 telescope of Special Astrophysical Observatory (SAO).

3.1   LIGO/Virgo G299232: Compact Binary Coalescence Candidate

G299232 is a low-significance compact binary coalescence candidate identified from LIGO Hanford Observatory (H1) and LIGO Livingston Observatory (L1) at 2017-08-25 13:13:31 UTC. If the candidate has astrophysical origin, it appears consistent with the NS-BH merger [48]. Subsequently, the event was not confirmed.

Localization map generated by the BAYESTAR pipelines [39] including data from H1, L1, and V1 detectors is presented in Fig. 2. The 90% credible region spans about 2040 deg^2. The a posteriori luminosity distance estimate is 339 ± 110 Mpc [48].

The IceCube Neutrino Observatory (a cubic-kilometer neutrino detector operating at the geographic South Pole, Antarctica) searched for track-like neutrino candidates (GFU) detected in a $[-500, 500]$ s interval since the LIGO/Virgo trigger G299232 [4]. Comparison of the candidate source directions of 7 temporally-coincident neutrinos to the BAYESTAR skymap is presented in Fig. 2.

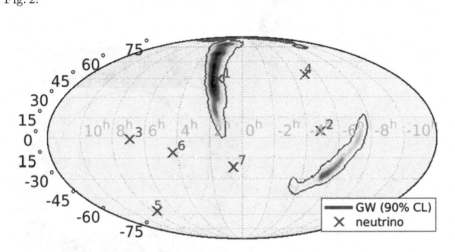

Fig. 2. The localization map generated by the BAYESTAR pipeline [39] based on data of H1, L1, and V1 detectors. X1–X7 are neutrino candidates (GFU) detected in a $[-500, 500]$ s interval since the LIGO-Virgo trigger G299232.

One of the neutrino candidates (marked as X1) was within the LIGO/Virgo localization area and registered 233.82 s before LIGO/Virgo triggered. X1 sky location is R.A. = 28.2, Dec. = 44.8 with 3.8° uncertainty of direction reconstruction [5].

We observed the field of LIGO/Virgo trigger G299232 [48] and error circle of IceCube candidate X1 [4,5] with wide field of view VT-78a telescope of ISON-Khureltogoot observatory. We obtained several unfiltered images with the two time series starting on 2017-08-25 (UT) 15:24:13 and 16:32:52 (time since LVC trigger are 0.11289 and 0.16054 days), each centered to the position of localization reported in [4] and [5], respectively. Total coverage of the error region of IceCube candidate X1 [5] is 85.7%. The map of the coverage can be found in Fig. 3.

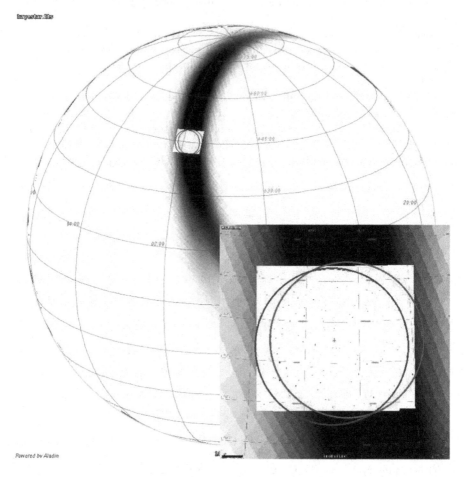

Fig. 3. The map of the coverage IceCube candidate X1 localization by VT-78a telescope of Khureltogoot observatory. Red circle is preliminary IceCube X1 error box [4], blue circle is final error box [5]. (Color figure online)

Using the procedure described in Sect. 2 we selected 94.7 thousand objects from the images (field of view is 7 × 7°). After comparing these 94.7 thousand objects with the USNO-B.1 catalog we have 834 candidates left, of which 818 are processing artifacts. Finally, we found one cataloged asteroid (895) Helio. We found no significant variability of the sources between the two epochs within one hour. We also did not find sources, which are weaker in the USNO-B1.0 catalog, representing potential host galaxies. Upper limit on the stellar magnitude of possible optical candidate is 19.2^m.

Assuming that the kilonova was exactly the same as in the case of GW170817, but placed at the distance of 339 ± 110 Mpc, the brightness of the kilonova at maximum in the R filter would be about 21^m. Our upper limit, which we obtained (19.2^m), is about 2 magnitudes worse, which is in agreement with results of our analysis.

3.2   Observations of LIGO/Virgo Optical Candidates

In addition to searching the object in the localization area, it is also important to provide follow up photometric and spectroscopic observations of previously found transients to identify the nature of them. In case of binary neutron stars or neutron star and black hole merging a type I (short) gamma-ray burst is expected, accompanied by an afterglow and kilonova emission.

We also observed objects in the localization area of GW events that were found by other research groups. The objects are listed in the Table 2, the localization maps of each gravitational-wave event are presented in the Fig. 4. Several gravitational-wave events were not officially confirmed later and remained as candidates.

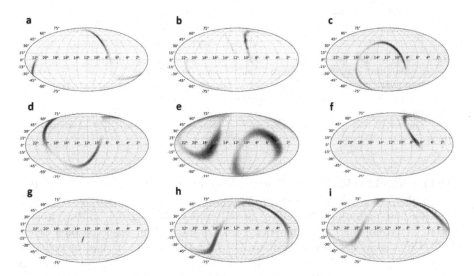

Fig. 4. Sky localization of LIGO/Virgo events. a—GW170104_G268556, b—GW170120_G270580,   c—GW170217_G274296,   d—GW170227_G275697,   e—GW170313_G277583,   f—GW170608_G288732,   g—GW170817_G298048,   h—GW170823_G298936, i—GW170825_G299232

Table 2. Observations of optical candidates of LIGO/Virgo events

Event	Type	Optical candidates	Type
GW170104_G268556	CBC (+) [43]	PS17dp	n/c
		PS17fl	n/c
		PS17fn	n/c
		PS17gl	n/c
GW170120_G270580	n/c (−) [44]	PS17lk	n/c
		PS17nv	n/c
		PS17pv	n/c
		PS17qk	n/c
		PS17rc	n/c
		PS17yt	SN Ia
		MASTER OT J090737.22+611200.5	n/c
GW170217_G274296	n/c (−) [45]	PS17bek	SLSN
GW170227_G275697	CBC (−) [46]	iPTF17bub	M-d.f.
		iPTF17bue	SN Ia
		XRT23	n/c
GW170313_G277583	n/c (−) [47]	ATLAS17cgg	SN IIn
GW170608_G288732	BBH (+) [3]	GW170608X2	n/c
GW170817_G298048	BNS (+) [3]	GW 170817	GRB, KN
GW170823_G298936	BBH (+) [3]	GWFUNC-17ure	SN Ia
GW170825_G299232	NS+BH (−) [48]	SwiftJ014008.5+343403.6	n/c
		MASTER OT J033744.97+723159.0	SN IIb

BBH – binary black holes merging, BNS – binary neutron stars merging, NS+BH – neutron star and black hole merging, CBC – compact binary coalescence, n/c – this event candidate does not have a chirp signature, and thus does not suggest a compact binary merger or the morphology of the event candidate is unclear.
(+) – event, (−) – candidate.
SN – supernova, KN – kilonova, SLSN – super-luminous supernova, GRB – gamma-ray burst, M-d.f. – M-dwarf flare, n/c – non classified.

GW170104_G268556. The Blanco telescope at CTIO observed about 10% (230 sq. deg.) of the localization probability map from 15 to 19 h after the GW detection and 77 optical candidates were found up to 5-sigma limiting magnitude of i∼23 AB mag [41]. The ATLAS telescope system observed part of the LIGO/Virgo map between 30 min to 3 h after the detection time of G268556 and 3 new optical candidates were found up to 5-sigma limiting magnitude of 19.0 mag in two filters ATLAS-cyan and ATLAS-orange (covering 420–650 and 560–820 nm, respectively) [50,51]. The Global MASTER Net found 21 optical candidates [30,31,52], 18 optical candidates were found with the TAROT network of telescopes [9], 56 optical candidates with R magnitudes from 16.6^m to 20.7^m were found by iPTF with the Palomar 48-in. Oschin telescope [24,38] and 49 unique optical candidates were found with the Pan-STARRS1 telescope [40].

We observed the following Pan-STARRS optical candidates: PS17dp with R.A. 09:00:51.09 Dec. +45:26:44.3, PS17fl with R.A. 08:02:09.73 Dec. +28:49:44.7, PS17fn with R.A. 08:05:51.40 Dec. +28:42:51.9, PS17gl with R.A. 08:34:50.85 Dec. +61:13:02.2. Results of our observations are presented in Table 3.

Table 3. The photometric observation results of orphan sources detected in the localization area of LIGO/Virgo event GW170104 (trigger G268556).

Orphan	Filter	MJD	Magnitude	Orphan	Filter	MJD	Magnitude
PS17dp	R	57770.77716	16.93 ± 0.04	PS17fn	R	57770.75515	19.51 ± 0.16
	R	57774.75795	16.91 ± 0.05		R	57772.74354	19.18 ± 0.04
	R	57775.7468	16.90 ± 0.05		R	57774.68807	19.11 ± 0.04
	R	57776.72092	16.93 ± 0.05		R	57775.69088	19.15 ± 0.04
	R	57777.89783	17.02 ± 0.04		R	57776.6997	19.15 ± 0.05
PS17fl	R	57770.69727	18.15 ± 0.10		R	57777.84441	19.31 ± 0.05
	R	57772.7159	18.16 ± 0.02		R	57778.71114	19.33 ± 0.04
	R	57774.65356	18.20 ± 0.02		R	57779.71823	18.99 ± 0.05
	R	57775.66825	18.18 ± 0.02		R	57781.80806	19.31 ± 0.07
	R	57776.66306	18.17 ± 0.02	PS17gl	R	57772.80185	17.38 ± 0.03
	R	57777.82865	18.21 ± 0.03		R	57774.71087	17.39 ± 0.03
	R	57778.69277	18.18 ± 0.03		R	57775.71707	17.40 ± 0.03
	R	57781.78667	18.12 ± 0.02		R	57777.87095	17.56 ± 0.02
	R	57840.73404	18.22 ± 0.03		R	57778.79509	17.57 ± 0.02

GW170120_G270580. The Pan-STARRS covered northern area of the GW localization and detected 124 transients including rapidly rising transient – PS17yt with R.A. 10:03:57.96 Dec. +49:02:28.3 [12,21]. Our collaboration observed PS17yt source in BVR filters and light curves were constructed (see Fig. 5a). It was subsequently shown that PS17yt is Ia type supernova at a redshift of z ~ 0.026 [10].

Furthermore, we observed the following orphan sources: PS17lk with R.A. 09:29:58.27 Dec. +15:11:58.5, PS17nv with R.A. 09:57:41.01 Dec. +17:49:33.4, PS17qk with R.A. 09:29:12.15 Dec. +25:49:06.4, PS17pv with R.A. 09:25:07.35 Dec. +50:12:28.9, PS17rc with R.A. 09:32:19.16 Dec. +47:03:38.3 and MASTER J090737.22+611200.5 with R.A. 09:07:37.22 Dec. +61:12:00.5 in the field of the LIGO G270580 localizations. Results of observations are presented in Table 4, 5.

GW170217_G274296. Pan-STARRS covered 501 square degrees of localization area on the first night following the release of the G274296 alert. They have located and vetted 10 transients with host spectroscopic redshifts and 60 unknown transients with no host spectroscopic redshifts [11]. We observed

Table 4. The photometric observation results of orphan source PS17yt detected in the localization area of LIGO/Virgo candidate GW170120 (trigger G270580).

Orphan	Filter	MJD	Magnitude	Filter	MJD	Magnitude
PS17yt	R	57777.92649	18.64 ± 0.03	R	57825.60539	20.28 ± 0.18
	Clear	57777.73377	18.49 ± 0.02	R	57826.91885	19.66 ± 0.25
	R	57779.71072	18.35 ± 0.03	Clear	57834.84494	19.75 ± 0.07
	R	57780.85639	18.14 ± 0.02	R	57841.93598	20.10 ± 0.04
	R	57781.83043	18.04 ± 0.02	R	57847.69544	19.79 ± 0.11
	R	57782.80977	18.11 ± 0.01	R	57859.74791	19.80 ± 0.06
	R	57782.85949	17.97 ± 0.01	R	57868.79662	20.77 ± 0.15
	R	57783.83928	18.04 ± 0.01	R	57868.99791	20.37 ± 0.04
	R	57784.09742	18.06 ± 0.01	R	57902.72538	20.83 ± 0.06
	R	57784.78692	18.01 ± 0.01	R	57902.81506	21.52 ± 0.25
	R	57785.83609	17.99 ± 0.01	B	57779.69005	18.42 ± 0.02
	R	57789.73838	17.94 ± 0.03	B	57784.11202	18.25 ± 0.01
	R	57790.8769	18.01 ± 0.03	B	57789.74278	18.09 ± 0.05
	R	57791.75238	18.03 ± 0.04	B	57791.78446	18.49 ± 0.07
	R	57791.71372	18.05 ± 0.03	B	57791.75104	18.29 ± 0.06
	R	57792.72201	18.06 ± 0.03	B	57792.7409	18.40 ± 0.06
	R	57793.9065	18.10 ± 0.04	B	57793.90649	18.37 ± 0.08
	Clear	57794.85808	18.10 ± 0.05	B	57795.82528	17.80 ± 0.14
	R	57797.73505	18.56 ± 0.02	B	57796.73038	18.29 ± 0.12
	R	57799.66676	18.34 ± 0.03	B	57797.72868	18.83 ± 0.06
	R	57799.97846	18.39 ± 0.05	B	57799.69759	18.97 ± 0.05
	R	57801.68115	18.52 ± 0.07	B	57799.9764	18.76 ± 0.07
	R	57802.81726	18.60 ± 0.03	B	57801.69569	18.98 ± 0.11
	r	57806.10869	18.84 ± 0.06	B	57802.78863	19.31 ± 0.04
	R	57806.81488	18.59 ± 0.06	B	57806.81282	19.49 ± 0.12
	R	57806.82609	18.66 ± 0.03	B	57807.80606	19.86 ± 0.09
	R	57807.80814	18.81 ± 0.05	B	57808.66562	19.93 ± 0.04
	R	57808.63134	18.70 ± 0.03	B	57808.97479	19.59 ± 0.07
	Clear	57808.73316	18.42 ± 0.05	B	57809.78226	19.98 ± 0.04
	R	57808.96605	18.89 ± 0.02	B	57810.01513	19.61 ± 0.02
	Clear	57809.88016	18.59 ± 0.07	B	57810.77539	19.75 ± 0.15
	R	57809.75215	18.72 ± 0.02	B	57811.85932	20.44 ± 0.19
	R	57810.00647	18.89 ± 0.01	B	57812.88898	20.23 ± 0.04
	R	57810.77711	18.60 ± 0.05	B	57819.75285	21.06 ± 0.16
	R	57811.7597	18.62 ± 0.05	B	57819.85285	20.55 ± 0.08
	R	57811.85895	19.05 ± 0.05	B	57820.68824	21.02 ± 0.35
	R	57811.88851	19.09 ± 0.02	B	57820.76758	20.50 ± 0.15
	R	57812.80762	18.79 ± 0.06	B	57824.83462	>19.4
	R	57816.76485	19.06 ± 0.04	B	57826.89029	>18.52
	Clear	57818.91924	19.12 ± 0.06	B	57841.94189	21.65 ± 0.08
	R	57819.75292	19.01 ± 0.08	B	57868.79491	>22.3
	R	57819.87185	19.08 ± 0.03	B	57869.01979	22.43 ± 0.16
	R	57820.68824	19.47 ± 0.14	V	57795.79292	17.88 ± 0.18
	R	57820.80353	18.98 ± 0.12	V	57796.69757	18.25 ± 0.1
	Clear	57821.92038	18.99 ± 0.2	V	57799.64108	18.46 ± 0.02
	R	57824.83463	19.16 ± 0.1	V	57801.65545	18.66 ± 0.05

Table 5. The photometric observation results of orphan sources detected in the localization area of LIGO/Virgo candidate GW170120 (trigger G270580).

Orphan	Filter	MJD	Magnitude	Orphan	Filter	MJD	Magnitude
PS17lk	R	57778.72969	20.92 ± 0.16	PS17nv	R	57778.74557	>22.2
	R	57782.74483	>22.5		R	57780.82339	>22.2
	R	57783.70939	21.10 ± 0.11		R	57784.03700	>23.4
	R	57784.69874	21.22 ± 0.12	PS17pv	R	57778.82847	>20.5
	R	57791.93095	21.60 ± 0.40		R	57781.86958	>22.4
	R	57802.73167	21.89 ± 0.24		R	57783.80134	20.74 ± 0.11
PS17qk	R	57778.77275	21.01 ± 0.12	PS17rc	R	57778.81470	20.96 ± 0.15
	R	57782.78859	20.61 ± 0.11		R	57783.77229	21.04 ± 0.10
	R	57783.74301	20.46 ± 0.06		R	57784.76358	21.13 ± 0.11
	R	57784.73083	20.36 ± 0.05		R	57784.88885	21.28 ± 0.05
	R	57784.84479	20.50 ± 0.03		B	57784.87345	23.57 ± 0.24
	B	57784.85668	20.87 ± 0.04		R	57785.80742	21.28 ± 0.09
	R	57785.77138	20.17 ± 0.05	MASTER OT	R	57774.62327	19.07 ± 0.01
	R	57802.75716	20.55 ± 0.07		R	57775.64491	19.10 ± 0.02
	Clear	57818.86623	20.53 ± 0.09		R	57776.63227	19.18 ± 0.03

one of the transients with no host spectroscopic redshifts (PS17bek with R.A. 10:47:41.90 Dec. +26:50:06.0) and the light curves in BR-filters are presented in the Fig. 5b and in the Table 6. Afterwards a good correlation between PS17bek spectrum and the spectra of super-luminous supernovae (SLSNe type I) was found. In particular, a good match with the spectra of SN 2010gx at −5 days (before peak) if PS17bek is at the redshift of z ∼ 0.31 was found. The weak emission line at 6559.4 is consistent with [O III] 5007 at z = 0.31, and were also detected [O III] 4959 at a consistent redshift but lower significance [20].

GW170227_G275697. We observed following iPTF optical transient candidates: iPTF17bub with R.A. 05:42:06 Dec. +70:09:35.2 and iPTF17bue with R.A. 03:38:29.70 Dec. +71:24:27.9. Results of observations are presented in Table 6. We also have optical observations of an X-ray source (XRT23) detected by Swift/XRT in coordinates R.A. 21:33:14.38 Dec. +70:09:35.2. The XRT23 source is of interest since a power-law fit to the light curves gives a slope of about 0.9 [17] which is similar a slope of a gamma-ray optical afterglow. But after a few days of observation showed that the source flux is remaining approximately constant and is unlikely to be related to the GW event [18]. We observed the XRT23 source several times but had not detected the source until 21.6^m in R filter.

GW170313_G277583. We observed a cataloged galaxy (R.A. 08:03:55.01 Dec. +26:31:14.1) with a redshift of 0.02, which contained a transient source

Table 6. The photometric observation results of orphan sources detected in the localization area of LIGO/Virgo candidate GW170217_G268556, GW170227_G275697.

Orphan	Filter	MJD	Magnitude	Orphan	Filter	MJD	Magnitude
PS17bek	R	57806.64857	19.43 ± 0.07		B	57812.93517	19.88 ± 0.03
	R	57806.79869	19.50 ± 0.11		B	57815.82638	20.04 ± 0.07
	R	57807.78079	19.43 ± 0.06		B	57819.82803	20.15 ± 0.09
	R	57808.94310	19.60 ± 0.06		B	57820.74384	20.12 ± 0.15
	R	57809.72846	19.51 ± 0.03		B	57841.95726	22.11 ± 0.15
	Clear	57809.82931	19.43 ± 0.04		B	57866.78736	>21.6
	R	57809.98522	19.62 ± 0.04		B	57868.89165	>24.1
	R	57810.70502	19.24 ± 0.05	PS17bub	R	57815.73877	18.96 ± 0.04
	R	57810.80135	19.43 ± 0.13		R	57817.64625	18.76 ± 0.06
	R	57811.90788	19.22 ± 0.13		R	57818.73877	18.86 ± 0.08
	R	57812.96667	19.66 ± 0.05		R	57819.70281	18.64 ± 0.07
	R	57815.79957	19.65 ± 0.05		R	57819.72515	18.96 ± 0.06
	R	57819.80657	19.80 ± 0.08		R	57820.06315	19.01 ± 0.09
	R	57820.70020	19.79 ± 0.16		R	57820.66625	19.12 ± 0.11
	R	57835.88609	20.18 ± 0.10		R	57824.78685	18.90 ± 0.10
	R	57841.95141	20.63 ± 0.07		R	57825.55785	19.15 ± 0.13
	R	57859.71150	21.47 ± 0.19		R	57826.74134	19.14 ± 0.08
	R	57866.78978	21.15 ± 0.21	PS17bue	R	57815.71498	16.82 ± 0.05
	R	57868.86980	22.17 ± 0.06		R	57817.58241	17.24 ± 0.07
	R	57893.74805	22.77 ± 0.32		R	57818.75660	16.71 ± 0.11
	R	57906.72194	>19.7		R	57819.66831	17.18 ± 0.04
	B	57807.79971	19.58 ± 0.15		R	57820.58851	17.40 ± 0.07
	B	57808.77974	19.69 ± 0.09		R	57820.64728	17.22 ± 0.08
	B	57808.95185	19.79 ± 0.07		R	57824.75208	17.35 ± 0.18
	B	57809.99466	19.85 ± 0.04		R	57826.63920	17.10 ± 0.22
	B	57810.79963	19.38 ± 0.28				

(ATLAS17cgg). Analysis of the source spectrum suggests that it is a supernova Type IIn [8].

GW170817_G298048. On August 17, 2017 at 12:41:04 UTC the LIGO/Virgo gravitational-wave detectors observed a binary neutron star merging for the first time. Also a gamma-ray burst GRB 170817A was registered by Fermi/GBM ~1.7 s after the coalescence [1]. In the intersection area of localization maps constructed by LIGO/Virgo and gamma-ray detectors (Fermi/GBM, IPN) a bright optical transient AT2017gfo was discovered at 11 h after the merger by the One-Meter, Two Hemisphere team using the 1 m Swope Telescope. The optical transient (kilonova) was independently detected by multiple teams within an hour [2]. The discovery was followed by a broadband follow-up observations. Our results of AT2017gfo optical observations are published in [2,35].

Table 7. The photometric observation results of orphan sources detected in the localization area of LIGO/Virgo event GW170823_G298936.

Orphan	Filter	MJD	Magnitude	Filter	MJD	Magnitude
GWFUNC-17ure	R	57989.94354	17.88 ± 0.05	R	57990.97318	18.05 ± 0.04
	R	57989.94571	17.88 ± 0.03	R	57999.93592	18.16 ± 0.01
	R	57989.94785	17.95 ± 0.03	R	58004.88356	18.25 ± 0.03
	R	57989.95000	17.73 ± 0.03	R	58010.89630	18.51 ± 0.03
	R	57989.95215	17.93 ± 0.03	R	58012.89588	18.59 ± 0.03
	R	57989.95431	17.98 ± 0.03	R	58011.96662	18.53 ± 0.03
	R	57989.95646	18.00 ± 0.05	R	58013.96994	18.60 ± 0.03
	R	57989.95861	18.04 ± 0.05	R	58013.95318	18.62 ± 0.03
	R	57989.96078	17.93 ± 0.05	R	58013.67084	18.52 ± 0.03
	R	57989.96293	17.92 ± 0.06	R	58015.89279	18.63 ± 0.03
	R	57989.96508	17.87 ± 0.06	R	58023.98670	18.71 ± 0.03
	R	57989.96723	17.92 ± 0.06	R	58013.66668	18.52 ± 0.05
	R	57989.96938	17.95 ± 0.09	R	58013.66807	18.62 ± 0.04
	R	57989.97368	17.98 ± 0.04	R	58013.66946	18.44 ± 0.04
	R	57989.97583	18.04 ± 0.04	R	58013.67084	18.45 ± 0.04
	R	57989.97799	18.01 ± 0.06	R	58013.67223	18.52 ± 0.04
	R	57989.98014	17.97 ± 0.04	R	58013.67362	18.58 ± 0.05
	R	57989.98229	18.01 ± 0.04	R	58013.67501	18.58 ± 0.05
	R	57989.98444	17.88 ± 0.03	R	58144.55350	19.53 ± 0.13
	R	57989.98696	17.95 ± 0.04	B	57999.93254	19.97 ± 0.22
	R	57989.98911	18.06 ± 0.05	B	58004.88150	19.89 ± 0.12
	R	57990.95387	18.02 ± 0.04	B	58010.89388	19.98 ± 0.16
	R	57990.95602	17.89 ± 0.03	B	58012.88390	20.29 ± 0.07
	R	57990.95816	18.04 ± 0.03	B	58011.93990	20.30 ± 0.09
	R	57990.96030	18.00 ± 0.03	B	58013.98197	20.32 ± 0.06
	R	57990.96245	17.99 ± 0.04	B	58013.95146	20.15 ± 0.09
	R	57990.96459	17.94 ± 0.03	B	58013.68650	20.15 ± 0.08
	R	57990.96675	17.98 ± 0.03	B	58015.89038	20.37 ± 0.08
	R	57990.96889	17.95 ± 0.03	B	58024.00543	20.41 ± 0.05
	R	57990.97104	17.95 ± 0.03			

To summarize, the kilonova AT2017gfo along with gamma-ray burst GRB 170817A are the first electromagnetic counterparts of a gravitational wave event of a binary neutron star merging. The event confirmed the connection of short gamma-ray bursts with the merging of two neutron stars, predicted theoretically.

GW170823_G298936. We observed CNEOST (Chinese Near Earth Object Survey Telescope) optical candidate GWFUNC-17ure with R.A. 03:40:39.45 Dec. +40:33:22.9 [55]. Results of observations are presented in Table 7.

GW170825_G299232. Global MASTER robotic net discovered an optical transient source – MASTER OT J033744.97+723159.0 with R.A. 03:37:44.97 Dec. +72:31:59.0 [29]. Analysis of the source spectrum suggests that it is a supernova Type IIb [23]. The source observation with the RoboPolpolarimeter shown that the R-band fractional polarization of the source is $1.8 \pm 0.47\%$ [36]. Our observations of the source are shown in the Fig. 5c and in the Table 8.

Table 8. The photometric observation results of orphan sources detected in the localization area of LIGO/Virgo candidate GW170825_G299232.

Orphan	Filter	MJD	Magnitude	Filter	MJD	Magnitude
MASTER OT	R	57994.97791	15.84 ± 0.01	B	57994.99543	16.87 ± 0.01
	R	57997.97357	15.55 ± 0.01	B	57996.87182	16.64 ± 0.02
	R	57999.59874	15.44 ± 0.01	B	57997.97416	16.59 ± 0.01
	R	58000.63263	15.40 ± 0.01	B	57999.63320	16.63 ± 0.01
	R	58005.00409	15.39 ± 0.01	B	58000.64482	16.66 ± 0.02
	R	58007.00562	15.47 ± 0.01	B	58006.99930	17.05 ± 0.04
	R	58007.98446	15.53 ± 0.01	B	58007.99935	17.20 ± 0.02
	R	58008.99011	15.60 ± 0.01	B	58009.00108	17.33 ± 0.03
	R	58009.96519	15.66 ± 0.01	B	58009.97612	17.47 ± 0.03
	R	58010.85784	15.75 ± 0.02	B	58010.85774	17.47 ± 0.02
	R	58011.04979	15.75 ± 0.01	B	58011.03549	17.67 ± 0.01
	R	58012.03532	15.81 ± 0.01	B	58012.01566	17.79 ± 0.01
	R	58012.95882	15.86 ± 0.01	B	58014.00297	17.98 ± 0.01
	R	58013.99710	15.67 ± 0.01	B	58015.94982	18.25 ± 0.02
	R	58015.95224	16.03 ± 0.01	B	58016.99917	18.24 ± 0.03
	R	58016.98923	16.10 ± 0.01	B	58024.03907	18.53 ± 0.02
	R	58024.03079	16.40 ± 0.01	B	58042.99286	18.79 ± 0.08
	R	58042.98260	16.98 ± 0.02	B	58043.88464	18.84 ± 0.01
	R	58043.86788	16.53 ± 0.01	B	58045.76583	18.90 ± 0.01
	R	58045.75628	16.90 ± 0.01			

4   Summary

In 2017, coordinated hardworking of thousands of astronomers and other scientists around the world allowed to find and successfully observe the first electromagnetic counterpart of the gravitational wave event GW170817 of binary neutron star merging. The associated GRB 170817A and kilonova AT2017gfo were observed by hundreds of space and ground-based telescopes and observatories for the first time in all ranges of electromagnetic spectrum. The unprecedented collaboration allowed to obtain detailed properties of kilonova and to verify existing physical models of this phenomenon, which is not fully studied yet. At the

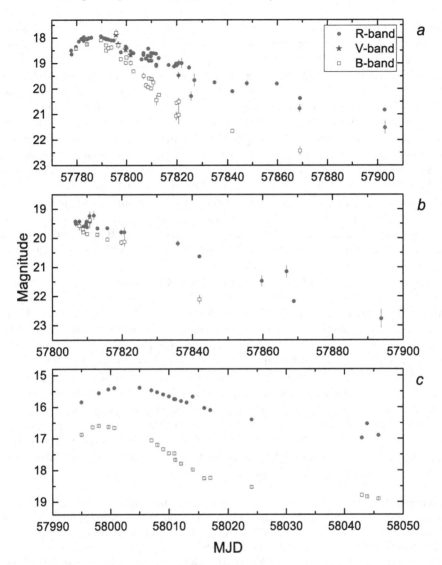

Fig. 5. (a) light curves of PS17yt (GW170120_G270580 optical candidate), (b) light curves of PS17bek (GW170217_G274296 optical candidate), (c) light curves of MASTER-OT (GW170825_G299232 optical candidate). Red circles points are R-band, blue square points are B-band and green star points are V-band. Host galaxy is not subtracted. (Color figure online)

same time, there was no any reliable EM counterpart candidate detected for 10 binary black holes coalescences discovered during O1/O2 scientific runs of LIGO and Virgo detectors. However, a huge amount of observational data, which covered vast localization area of the events, led to the discovery of many other new transient sources unrelated to the GW. The problem of search of a new optical transient with specific properties in large localization areas arose here with the great actuality.

We discussed the two main methods of the search for optical transients in the areas of tens and hundreds of square degrees: mosaic surveys and observations of pre-defined targets (potential host galaxies). The case of mosaic surveys is suitable for small-aperture telescopes with wide fields of view, with rather low optical upper limit, though. The search of the transient inside pre-defined target galaxies requires deeper limits and thus require observations with large-aperture telescopes with >1 m diameter. The second case involves compiled catalogues of galaxies with known distance like Galaxy List for the Advanced Detector Era (GLADE) [22]. This fact increases the actuality of deep surveys of galaxies with measured distances. These methods are suitable not only for the search of the EM counterpart of gravitational waves events detected by LIGO/Virgo, but also for the search of optical counterparts of ordinary GRBs with large localization region (e.g., from GBM/Fermi experiment).

We also provided results of the observations of localization regions of candidates for real GW events detected with LIGO/Virgo during their second scientific run O2. We did not find any new optical transients with our facilities; however, we conducted a follow-up of transients discovered by other teams worldwide. This valuable experience is now being adapted for the third scientific run O3 of LIGO/Virgo, which started on April 1, 2019 and would continue for 1 year. Nevertheless, the problem of automatization of the data processing algorithms remains unsolved for all cases and requires the development of new conceptual approach, and generalized pipelines for data reduction are required.

Almost all space and ground-based astronomical facilities are now involved in the follow-up of GW events. This makes multi-messenger astronomy a commonplace nowadays. Quick availability of new obtained data and vast collaboration of observatories and observers may guarantee further success.

Acknowledgments. EM, AP, AV, PM, SB thank the staff of all the observatories for providing data. The authors are grateful for a partial financial support to the grants RFBR 17-51-44018, 18-32-00784 and 19-42-910014 provided jointly by RFBF and the Republic of Crimea. Observations of AS32 telesopes of Abastumani Astrophysical Observatory are supported by Shota Rustaveli Science Foundation grant RF-18-1193. Observations of the Tian Shan Astrophysical Observatory are carried out in the framework of Project No. BR05236322 "Studies of physical processes in extragalactic and galactic objects and their subsystems", financed by the Ministry of Education and Science of the Republic of Kazakhstan.

References

1. Abbott, B.P., Abbott, R., Abbott, T.D., et al.: GW170817: observation of gravitational waves from a binary neutron star inspiral. Phys. Rev. Lett. **119**(16), 161101 (2017). https://doi.org/10.1103/PhysRevLett.119.161101

2. Abbott, B.P., Abbott, R., Abbott, T.D., et al.: Multi-messenger observations of a binary neutron star merger. Astrophys. J. Lett. **848**(2), L12 (2017). https://doi.org/10.3847/2041-8213/aa91c9

3. Abbott, B.P., Abbott, R., Abbott, T.D., et al.: GWTC-1: a gravitational-wave transient catalog of compact binary mergers observed by LIGO and Virgo during the first and second observing runs. Phys. Rev. X **9**(3), 031040 (2019). https://doi.org/10.1103/PhysRevX.9.031040

4. Bartos, I., Countryman, S., Finley, C., Blaufuss, E., Corley, R., Marka, Z.: LIGO/Virgo G299232: FOUND COINCIDENT IceCube neutrino observation. GRB Coordinates Network **21694**, 1, August 2017

5. Bartos, I., et al.: LIGO/Virgo G299232: COINCIDENT IceCube neutrino observation UPDATE. GRB Coordinates Network **21698**, 1, August 2017

6. Bertin, E., Arnouts, S.: SExtractor: software for source extraction. Astron. Astrophys. Suppl. **117**, 393–404 (1996). https://doi.org/10.1051/aas:1996164

7. Bisikalo, D.V., Zhilkin, A.G., Kurbatov, E.P.: Possible electromagnetic manifestations of merging black holes. Astron. Rep. **63**(1), 1–14 (2019). https://doi.org/10.1134/S1063772919010025

8. Blagorodnova, N., Burdge, K., Adams, S.M., Kasliwal, M.M.: LIGO/Virgo G277583: P200/DBSP classification of ATLAS candidates. GRB Coordinates Network **20897**, 1, March 2017

9. Boer, M., Laugier, R., Noysena, K., Klotz, A.: LIGO/Virgo G268556: TZAC TAROT network observations with TRE/TCA/TCH. GRB Coordinates Network **20445**, 1, January 2017

10. Castro-Tirado, A.J., Casanova, V., Zhang, B.B., et al.: LIGO/Virgo G270580: 10.4m GTC spectroscopic observations of PS17yt. GRB Coordinates Network 20521, 1, June 2017

11. Chambers, K.C., Smith, K.W., Huber, M.E., et al.: LIGO/Virgo G274296: Pan-STARRS imaging and discovery of 70 transients. GRB Coordinates Network **20699**, 1, June 2017

12. Chambers, K.C., Smith, K.W., Young, D.R., et al.: LIGO/Virgo G270580: Pan-STARRS coverage and bright, rising transient PS17yt. GRB Coordinates Network **20512**, 1, June 2017

13. Cook, D.O., van Sistine, A., Singer, L., Kasliwal, M.M.: LIGO/Virgo G298048: nearby galaxies in the localization volume. GRB Coordinates Network **21519**, 1, August 2017

14. Dálya, G., et al.: GLADE: a galaxy catalogue for multimessenger searches in the advanced gravitational-wave detector era. Mon. Not. R. Astron. Soc. **479**(2), 2374–2381 (2018). https://doi.org/10.1093/mnras/sty1703

15. Devyatkin, A.V., Gorshanov, D.L., Kouprianov, V.V., Verestchagina, I.A.: Apex I and Apex II software packages for the reduction of astronomical CCD observations. Sol. Syst. Res. **44**(1), 68–80 (2010). https://doi.org/10.1134/S0038094610010090

16. EHT: event horizon telescope. https://eventhorizontelescope.org. Accessed 30 Jan 2020

17. Evans, P., Breeveld, A., Kennea, J., et al.: LIGO/Virgo G275697: Swift-XRT source 23 is fading. GRB Coordinates Network **20841**, 1, March 2017

18. Evans, P., Kennea, J., Barthelmy, S., et al.: LIGO/Virgo G275697: Swift-XRT source 23 no longer fading. GRB Coordinates Network **20841**, 1, March 2017

19. Gaia: the most precise 3D space catalog. http://sci.esa.int/gaia. Accessed 30 Jan 2020

20. Gal-Yam, A., Leloudas, G., Vreeswijk (Weizmann), P.E., et al.: LIGO/Virgo G274296: PS17bek is a superluminous supernova at z=0.31. GRB Coordinates Network **20721**, 1, February 2017

21. Huber, M.E., Chambers, K.C., Smith, K.W., et al.: LIGO/Virgo G270580: Pan-STARRS coverage and 124 optical transients. GRB Coordinates Network **20518**, 1, June 2017

22. IceCUBE: south pole neutrino observatory. https://icecube.wisc.edu. Accessed 30 Jan 2020

23. Jonker, P., Fraser, M., Nissanke, S., et al.: LIGO/Virgo G299232: WHT spectrum of MASTER OT J033744.97+723159.0. GRB Coordinates Network **21737**, 1, August 2017

24. Kasliwal, M.M., Singer, L.P., Karamehmetoglu, E., et al.: IGO/Virgo G268556: iPTF optical transient candidates. GRB Coordinates Network **20398**, 1, January 2017

25. Lang, D., Hogg, D.W., Mierle, K., Blanton, M., Roweis, S.: Astrometry.net: blind astrometric calibration of arbitrary astronomical images. Astron. J. **139**(5), 1782–1800 (2010). https://doi.org/10.1088/0004-6256/139/5/1782

26. Li, L.X., Paczyński, B.: Transient events from neutron star mergers. Astrophys. J. **507**(1), L59–L62 (1998). https://doi.org/10.1086/311680

27. LIGO: the laser interferometer gravitational-wave observatory. https://www.ligo.caltech.edu. Accessed 30 Jan 2020

28. LIGO Scientific Collaboration, Aasi, J., Abbott, B.P., Abbott, R., Abbott, T., Abernathy, M.R., et al.: Advanced LIGO. Class. Quantum Gravity **32**(7), 074001 (2015). https://doi.org/10.1088/0264-9381/32/7/074001

29. Lipunov, V., Gorbovskoy, E., Kornilov, V., et al.: LIGO/Virgo G299232/PGWB170825.55: MASTER Global-Net OT inside NGC1343 discovery. GRB Coordinates Network **21719**, 1, August 2017

30. Lipunov, V., Gress, O., Tyurina, N., et al.: LIGO/Virgo G268556/GW170104: global MASTER Net second OT detection. GRB Coordinates Network **20392**, 1, January 2017

31. Lipunov, V., Tyurina, N., Gorbovskoy, E., et al.: LIGO/Virgo G268556/GW170104: global MASTER Net observations and first OT detection. GRB Coordinates Network **20381**, 1, January 2017

32. LSST: large synoptic survey telescope. https://www.lsst.org/. Accessed 30 Jan 2020

33. Mazaeva, E., Pozanenko, A., Volnova, A., et al.: Search and observations of optical counterparts for events registered by LIGO/Virgo gravitational wave detectors. In: CEUR Workshop Proceedings, vol. 2523, no. 1, pp. 229–243 (2019)

34. Meegan, C., et al.: The Fermi gamma-ray burst monitor. Astrophys. J. **702**(1), 791–804 (2009). https://doi.org/10.1088/0004-637X/702/1/791

35. Pozanenko, A.S., et al.: GRB 170817A associated with GW170817: multi-frequency observations and modeling of prompt gamma-ray emission. Astrophys. J. Lett. **852**(2), L30 (2018). https://doi.org/10.3847/2041-8213/aaa2f6

36. Reig, P., Panopoulou, G.: LIGO/Virgo G299232: RoboPol observations of MASTER OT J033744.97+723159.0. GRB Coordinates Network **21802**, 1, August 2017

37. SDSS: the sloan digital sky survey. https://www.sdss.org. Accessed 30 Jan 2020

38. Singer, L.P., Kupfer, T., Roy, R., et al.: LIGO/Virgo G268556: continued iPTF observations and additional optical transient candidates. GRB Coordinates Network **20401**, 1, January 2017

39. Singer, L.P., et al.: Going the distance: mapping host galaxies of LIGO and Virgo sources in three dimensions using local cosmography and targeted follow-up. Astrophys. J. Lett. **829**(1), L15 (2016). https://doi.org/10.3847/2041-8205/829/1/L15

40. Smartt, S.J., Smith, K.W., Huber, M.E., et al.: LIGO/Virgo G268556: Pan-STARRS1 observations and summary of transient sources. GRB Coordinates Network **20401**, 1, January 2017

41. Soares-Santos, M., Annis, J., Berger, E., et al.: LIGO/Virgo G268556: DESGW/DECam observations. GRB Coordinates Network **20376**, 1, January 2017

42. Tanvir, N.R., Levan, A.J., González-Fernández, C., et al.: The emergence of a Lanthanide-rich Kilonova following the merger of two neutron stars. Astrophys. J. Lett. **848**(2), L27 (2017). https://doi.org/10.3847/2041-8213/aa90b6

43. The LIGO Scientific Collaboration and the Virgo Collaboration: LIGO/Virgo G268556: updated sky map from gravitational-wave data. GRB Coordinates Network **20385**, 1, January 2017

44. The LIGO Scientific Collaboration and the Virgo Collaboration: LIGO/Virgo G270580: identification of a GW burst candidate. GRB Coordinates Network **20486**, 1, January 2017

45. The LIGO Scientific Collaboration and the Virgo Collaboration: LIGO/Virgo G274296: identification of a GW burst candidate. GRB Coordinates Network **20689**, 1, February 2017

46. The LIGO Scientific Collaboration and the Virgo Collaboration: LIGO/Virgo G275697: updated localization from LIGO data. GRB Coordinates Network **20833**, 1, March 2017

47. The LIGO Scientific Collaboration and the Virgo Collaboration: LIGO/Virgo G277583: identification of a GW burst candidate. GRB Coordinates Network **20860**, 1, March 2017

48. The LIGO Scientific Collaboration and the Virgo Collaboration: LIGO/Virgo G299232: identification of a GW compact binary coalescence candidate. GRB Coordinates Network **21693**, 1, August 2017

49. The LIGO Scientific Collaboration and the Virgo Collaboration: LIGO/Virgo identification of a binary neutron star candidate coincident with Fermi GBM trigger 524666471/170817529. GRB Coordinates Network **21509**, 1, August 2017

50. Tonry, J., Denneau, L., Heinze, A., et al.: LIGO/Virgo G268556: ATLAS imaging of the skymap. GRB Coordinates Network **20377**, 1, January 2017

51. Tonry, J., Denneau, L., Heinze, A., et al.: SUBJECT: LIGO/Virgo G268556: ATLAS17aeu - an unusual transient within the skymap. GRB Coordinates Network **20382**, 1, January 2017

52. Tyurina, N., Lipunov, V., Gress, O., et al.: LIGO/Virgo G268556: global MASTER Net 21 OTs detection. GRB Coordinates Network **20493**, 1, January 2017

53. Valenti, S., et al.: The discovery of the electromagnetic counterpart of GW170817: kilonova AT 2017gfo/DLT17ck. Astrophys. J. Lett. **848**(2), L24 (2017). https://doi.org/10.3847/2041-8213/aa8edf

54. Virgo: the gravitational-wave observatory. www.virgo-gw.eu. Accessed 30 Jan 2020

55. Xu, D., Li, B., Zhao, H., et al.: LIGO/Virgo G298936: GWFUNC/CNEOST optical observations and an optical transient. GRB Coordinates Network **21675**, 1, August 2017

Information Extraction from Text

Google Books Ngram: Problems
of Representativeness and Data Reliability

Valery D. Solovyev[1][(⊠)] [ID], Vladimir V. Bochkarev[1] [ID],
and Svetlana S. Akhtyamova[2] [ID]

[1] Kazan Federal University, Kazan, Russia
Maki.solovyev@mail.ru
[2] Kazan National Research Technological University, Kazan, Russia

Abstract. The article discusses representativeness of Google Books Ngram as a multi-purpose corpus. Criticism of the corpus is analysed and discussed. A comparative study of the GBN data and the data obtained using the Russian National Corpus and the General Internet Corpus of Russian is performed to show that the Google Books Ngram corpus can be successfully used for corpus-based studies. A new concept "diachronically balanced corpus" is introduced. Besides, the article describes the problems of word spelling and metadata errors presented in the GBN corpus and proposes possible ways of improving quality of the GBN data.

Keywords: Text corpora · Google Books Ngram · Corpus representativeness · Word frequency

1 Introduction

Corpus linguistics deals with machine-readable texts making text corpora an indispensable source of linguistic evidence. A corpus approach has proved to be highly useful in many areas of linguistics, such as theoretical linguistics, sociolinguistics, psycholinguistics, discourse analysis, semantics, contrastive linguistics, lexicography, and mathematical linguistics.

There are various types of text corpora and methods of their creation. They depend on specific research goals and are classified according to different criteria. However, one the most important features of any corpus is its representativeness (or balance). There are different definitions of corpus representativeness. Some scientists believe that to be representative and used for research purposes, a text corpus should include enough word forms and texts of different genres, periods, styles etc. distributed in a certain proportion within the corpus. Such definition can be applied to multi-purpose corpora which are used as a small model of a language reflecting the language life in a variety of its forms. However, there are corpora created for special purposes. Its representativeness depends on how well and objective the corpus reflects the investigated phenomenon [1].

There are corpora regarded to be representative in terms of balance and amount of data. One of them is the Russian National Corpus [2] (abbreviated as RNC). It is a well-known Russian research corpus detailed described in [3, 4]. There are more than

© Springer Nature Switzerland AG 2020
A. Elizarov et al. (Eds.): DAMDID/RCDL 2019, CCIS 1223, pp. 147–162, 2020.
https://doi.org/10.1007/978-3-030-51913-1_10

350 million of words in RNC [2]. It contains texts of different genres representing written Russian language. The corpus texts are selected and processed by professional linguists. Some of the homonyms are disambiguated manually. These factors make RNC an exceptionally good Russian language database which can be used in various linguistic studies.

The other Russian corpus considered to be representative is the General Internet-Corpus of Russian (http://www.webcorpora.ru/, abbreviated as GICR) [5]. It contains more than 20 billion of words and is expected to be updated. GICR includes texts from the largest Runet resources and social networks: Zhurnal'nyi Zal, Novosti, VKontakte, LiveJournal, and Blogs at Mail.ru. The corpora metadata include information on text genre, author, time of publication etc. It can be successfully used to solve problems of modern linguistics.

The largest Russian corpus is the Russian subcorpus of Google Books Ngram (https://books.google.com/ngrams, abbreviated as GBN). It contains more than 67 billion words (200 times as much as RNC). Besides Russian, it contains data on 7 languages. For example, the English subcorpus has 470 billion words. GBN is a database of scanned and recognized texts taken from the books of the 40 largest libraries of the world such as the libraries of Harvard University and Oxford University. Having digitized 30 million of books, 8 million best digitalized ones were selected to create the corpus (which is 6% of all the books published worldwide) [6]. A detailed description of the GBN can be found in [6–8]. Some scientists have negative attitude to GBN claiming that it contains a lot of ORC and metadata errors [9–11] and is unbalanced.

However, the GBN corpus is the largest of all the existing text corpora which makes it a unique database for scientific studies. The objective of this paper is to consider the GBN corpus representativeness paying special attention to the GBN problems mentioned in some scientific papers. We perform analysis of the problems and offer possible ways of making research work with the GBN data more reliable.

This article is an extended version of [12] which was significantly modified. Calculation results of ORC error probability were added (see Sect. 2, Fig. 2, 3 and Appendix). The data on frequency of words written according to the pre-reform spelling rules were added (see Sect. 2). *Introduction* was modified and *Conclusion* was rewritten. Literature review was extended. Some changes were also introduced in the other sections.

2   OCR Errors

It is a known fact that the GBN corpus still contains recognition errors. Most of them are due to poor print quality of the ancient books. There have been two versions of the GBN corpus. The first one was released in 2009 and contained a lot of errors. For example, the letter s was frequently recognized as f in ancient English books; and the word *best* was incorrectly recognized as *beft* in approximately half of the scanned books written in the 17th century. Google considered the criticism and significantly improved the recognition quality. The scanning devices were upgraded every six months [8]. The second updated version of the GBN corpus was released in 2012. In this version, the erroneous recognition of the word *best* (as *beft*) was only 10% and associated with the books written in the 17th century. As for contemporary books written in the 21st century, the error percentage was 0.02%. The percentage is rather low and cannot be statistically significant for frequency analysis of the word *best*.

The Russian subcorpus of GBN also contains recognition errors. For example, the letter *н* is sometimes recognized as *и*. Figure 1 shows frequency diagrams of the word *иней* (hoarfrost) used in GBN. In total, the word *иней* was used 78,148 times and the word *ииеü* (incorrect spelling) was used 303 times in 1920–2009. Thus, the probability of incorrect recognition in this case was 0.39%.

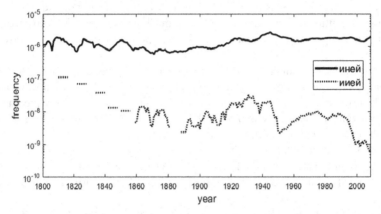

Fig. 1. Frequency of the words *иней* and *ииеü*

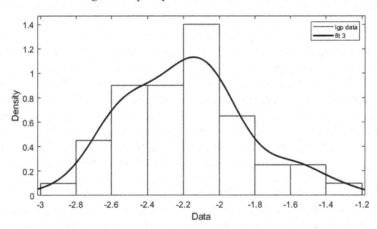

Fig. 2. The distribution of the decimal logarithm of the recognition error relative frequency. The solid line shows the kernel-smoothing estimation of the distribution density for the decimal logarithm of the error frequency

To estimate the probability of errors in a typical case, 100 of the most frequent incorrectly recognized word forms were selected. To do this, word forms found in the GICR [5] corpus and in the Open-Corpora electronic thesaurus [13, 14] were excluded from the list of word forms (sorted in descending order) found in the Google Books Ngram corpus. One hundred most frequent word forms with recognition errors were manually selected from the top of the obtained list of words. The most frequent errors are the following: 1) the letter combinations *ии, ий* and *ни* are erroneously recognized as the letter *ш* (50 cases in the obtained list); 2) the letters *и* and *н* are confused (48 cases in the obtained list). As for the rest two cases, the letter combination *ци* was incorrectly recognized as *щ* (*революции* as *револющи*) and the letter combination *ик* was erroneously recognized as the letter *ж* (*республики* as *республжи*). The full list of 100 words with the ORC errors is given in *Appendix*. The frequency of correctly and erroneously recognized variants was calculated for each case. The relative frequency of erroneous recognition was estimated. The distribution of the relative frequency of erroneous recognition is shown in Fig. 2. The decimal logarithms of the relative frequency are plotted horizontally. Thus, a value of −2 corresponds to 1% of cases of erroneous recognition.

The error relative frequency is less than 1% in 75% of cases from the selected list. The average value of the error relative frequency was 0.91% The median value may be more indicative since it shows behavior of a quantity in a typical case. The median value of the error relative frequency is 0.68%. Thus, frequency distortions are insignificant even in the cases of the most frequent recognition errors.

Figure 3 shows the change in the relative frequency of recognition errors for 100 selected examples. The results for 1- and 2-letter errors are shown separately. It is seen that frequency distortions are small after the 20 s of the 20th century and they have fallen significantly in recent decades.

There are certain difficulties associated with the pre-reform (before 1918) Russian orthography. Two letters Ѣ (yat) and Ѳ (fita) were used in Russian before the spelling reform of 1918. They are incorrectly recognized in GBN books written before 1918.

The words having the ending letter Ъ (er) are found more often. This letter is recognized in the corpus as the hard sign belonging to the alphabet established after the spelling reform. This happens because these letters resemble each other. In total, there are 269.658 Russian 1-grams with the ending letter Ъ. There are 168.307 1-grams in this list that consist only of the alphabet letters and have analogous forms which differs from them only by the presence of the letter Ъ. For such cases, it is obvious that the presence of the letter Ъ is the only orthographical difference before and after 1918. The most frequent 1000 pairs of words with and without the ending letter Ъ were selected. For each pair, the percentage of each variant was calculated in the total frequency of use for each year. Figure 4 shows the change of the median percentage of the variants for the given sample.

A number of words written according to the former spelling rules (with the ending Ъ) can still be found in the texts published after 1918. This can be explained by the presence of reprinted books and scientific texts in the corpus. The corpus contains a lot of words having the ending letter Ъ. However, they can be easily found and equated with the new forms. Thus, their spelling cannot be regarded as a serious problem in the process of the language dynamics analysis.

The stated examples show that recognition errors can still be found in the GBN corpus. However, their percentage is relatively low, and they cannot seriously affect the results of statistical research.

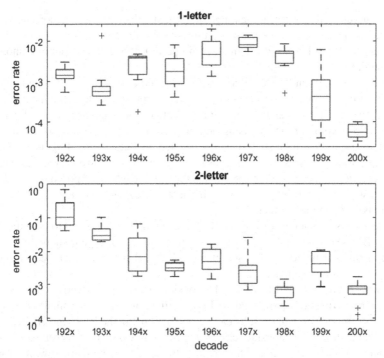

Fig. 3. Relative frequencies of 1- and 2-letter errors for the 100 most frequent incorrectly recognized word forms

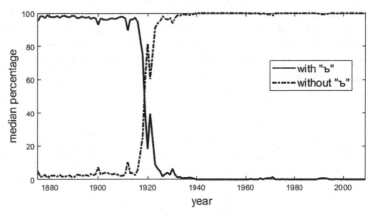

Fig. 4. The median percentage of the words with and without the ending ъ for the sample of 1000 most frequent word pairs

3   Representativeness (Balance)

Representativeness is a key issue in a corpus overall design. There are different approaches to defining this notion. However, the main requirement is that a corpus should relatively fully represent the range of linguistic distributions. According to [2], a representative corpus should contain certain proportion of texts of different genres and types which reflect a language life in a certain period. Besides, text samples should be long enough to obtain reliable data.

There is also a point of view that a well-balanced corpus is a utopia and any data obtained using a corpus reflect the content of that corpus rather than the language state [9, 10]. However, if this was true, creation of text corpora (requiring a lot of time and effort) would become an unpromising activity. In such case, corpora would just be a source of examples for linguists and cannot be used for fundamental linguistic research.

Compilation of a representative corpus is not an easy task. However, there are corpora that are regarded to be representative. For example, the RNC is supposed to be balanced because it was compiled by professional linguists. They sampled texts in the required proportions and annotated them.

There are also resources of linguistic evidence which are thought to be unbalanced such as GBN. GBN is a collection of tagged texts, an extra-large resource of scanned books. It is often criticized for being unbalanced [9, 10] and is not regarded as a corpus in terms of representativeness.

Anyway, both RNC and GBN can be considered as multi-purpose corpora aimed at performing various linguistic studies. In this article, we conduct a small comparative research to demonstrate what results can be obtained using the RNC and GBN corpora.

In [15], the task was to consider changes in frequency of two verbs *старался* (tried) and *пытался* (made an attempt) which belong to the same synonymic row. Both words are third person singular masculine verbs in the Past Tense. They turned to be the most frequently used verbs (in the RNC and GBN) from the corresponding inflectional paradigm. Frequency graphs were built for these words (see Figs. 5 and 6) and it was clearly seen that the word *пытался* becomes more frequent than the word *старался* in both corpora. The GBN graphs look smoother since the GBN corpus is larger than the RNC. The period when the word *пытался* becomes more frequent is the same for the studied corpora – approximately 1960. As for highly frequent words, the GBN and RNC frequency graphs are usually similar.

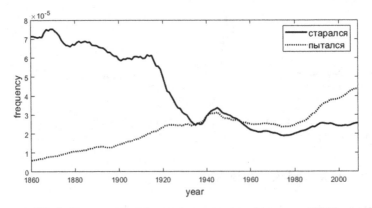

Fig. 5. Frequencies of the words *старался* and *пытался* (GBN)

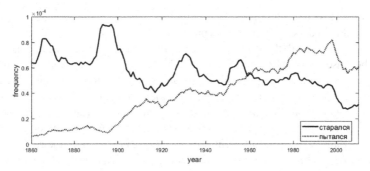

Fig. 6. Frequencies of the words *старался* and *пытался* (RNC)

Table 1 shows the GBN-based frequencies of the words *пытался* and *старался* in 1978 and in 2008, as well as their frequency ratio. Using the linear regression method, the expected frequency values were computed for 2014. This year was chosen to compare our predictions with the data obtained in [9], in which the time interval is limited to the period 2014–2015 because the data were available for the authors only before that time interval. The increased frequency of the word *пытался* compared to the frequency of the word *старался* within a 30-year period (1978–2008) allows us to expect its further growth by 2014.

Table 1. Known and predicted frequencies of the words *пытался* and *старался* in GBN

Word	1978	2008	Prediction for 2014
пытался	0.00173	0.00318	0.00347
старался	0.00140	0.00181	0.00189
пытался/старался	1.24	1.75	1.84

Let us compare the data obtained using GICR and described in [9] with the GBN frequency data. We will group and consider the GICR sub-corpora that differ in their genres and styles in the following way: 1) Zhurnalny Zal. It contains texts from literary journals. Its texts are similar to the GBN texts in terms of genre and style. 2) Novosti. This subcorpora contains texts of journalistic style. 3) LiveJournal and VKontakte. Both subcorpora contain texts that are different from book texts. Hence, one can expect that data obtained using the subcorpus Zhurnalny Zal will be similar to that of GBN and the data obtained using the other subcorpora will be different.

The frequency ratio of the word *пытался* to the word *старался* is 1.94 in Zhurnalny Zal, 10.41 in Novosti, and 3.30 in the subcorpora of the third group [9]. Thus, the ratio is different in the texts of different genres and styles which seems to be natural. However, if talking about texts of the same style and genre (such as the GBN books and literary journal texts), the values predicted based on the GBN corpus and the real values turned out to be very close to each other, differing by less than 6%. This indicates that reliable data can be obtained using the GBN corpus.

Unfortunately, not all the studies performed using GBN can be repeated using RNC or GICR because, unlike GBN, the RNC and GICR corpora are not available to users for downloading. This limits the possibilities of processing the RNC and GICR data with simple queries and does not allow applying complex computer-aided and mathematical data-processing methods that are widely used in contemporary research. The latter ones include measuring distances between languages at some point of time or between the states of a certain language at different time instants using different metrics [16].

In our opinion, the GBN text collection can be regarded as a representative written language corpus for the following reasons. The corpus overall size is extremely large and contains texts of optimal size. It includes scanned books taken from the largest world libraries. Thus, its texts are of various types and genres, written in different time and by authors of different age and sex. They represent the entire library system created by humans which cannot be achieved by manual text selection. The corpus size and diversity allow relatively reliable representation of distribution of linguistic features. As for the proportion of texts of different genres, we believe that it is not ideal but enough (taking into account the corpus size and origin) to make certain scientific conclusions.

Scientists developed sets of principles for achieving representativeness in a corpus design, for example they are detailly discussed in [17]. However, none of them mentioned diachronic balance, a term which is introduced in this paper.

By a "diachronically balanced" corpus we understand a corpus that is representative for any given moment of time, ideally for every year or decade. That is, any section of a corpus must be representative in any short time interval.

Until now, the problem of creating diachronically balanced corpora has not even been raised. However, the great size of the GBN corpus, as well as the adopted ideology of total scanning, make this corpus diachronically balanced.

The Russian subcorpus of GBN includes approximately 1 billion words per year over the past decades. It is triple as much as the volume of the entire RNC. The size of the English subcorpus per year is 10 times larger than the Russian one.

Let us consider a specific example demonstrating the degree of the GBN balance as compared to the RNC. In the late 1980s (the period of the USSR), the word *ускорение* (acceleration) used in physics began to be used in the political texts meaning the accelerated development of the country's economy. The wide use of this term is associated with M.S. Gorbachev's report at the Plenum of the Central Committee of the Communist Party of the USSR (CC CPSU) on the 23rd of April in 1985. He declared a large-scale reform program under the slogan of accelerating the social and economic development of the country. Two years later, in January 1987, at the Plenum of the CC CPSU, cardinal reconstructing of the economy management was announced. The new slogan of *перестройка* (reconstruction) appeared, and the frequency of the word *ускорение* began to decrease since it became less relevant. The graphs below show the frequency graphs of the word *ускорение* obtained using the GBN and RNC corpora.

Figure 7 shows that frequency of the word *ускорение* started growing rapidly exactly in 1985 and sharp decrease began in 1987. The GBN contains a lot of social-political materials written in the considered time. Thus, it is seen that the GBN data exactly reflect the social and political processes occurring at that time. It is not the single case. For example, it is shown in [18] that changes in languages registered in GBN correlate with social events. Let us consider the frequency of the word *ускорение* in the RNC corpus.

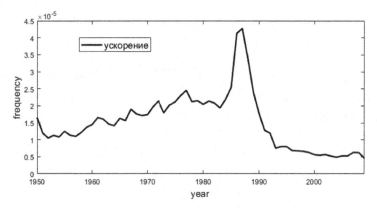

Fig. 7. Frequency of the word *ускорение* (GBN)

Figure 8 shows no growth in the frequency of the word *ускорение* in the RNC corpus in that period. On the contrary, the frequency of the word *ускорение* starts falling in 1985 and growing in 1987. There are no political texts among the specific ones containing the word *ускорение* in the RNC corpus in those years. This is just one example. However, it makes it clear how difficult it is to provide the diachronic balance when a corpus is compiled manually. Total digitalizing can contribute much to ensure this kind of balance.

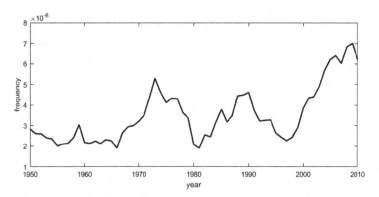

Fig. 8. Frequency of the word *ускорение* (RNC)

4   Errors in Metadata

There was a metadata error found in the English subcorpus *Fiction*. Many scientific books were mistakenly included into it in the first version of the GBN. This was found in [11] when the authors processed the *Fiction* subcorpus to compare the use of the word *Figure* typical of scientific text and the word *figure* (lowercased) that can also occur in fiction. Figure 9 shows an unnatural growth in frequency of the word *Figure* that corresponds to the time of the exponential growth of scientific publications.

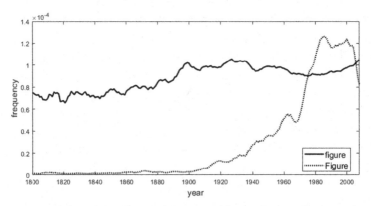

Fig. 9. Frequency of the words *Figure* and *figure* used in the *Fiction* subcorpus (the 1st GBN version)

This was considered in the second version of the GBN corpus and the books were classified correctly. Therefore, the frequency use of the word 'Figure' in the Fiction subcorpus decreased 20 times (Fig. 10). In that case, Google also rapidly responded to the criticism and the error was corrected.

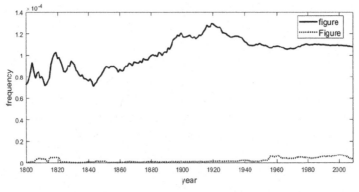

Fig. 10. Frequencies of the words *Figure* and *figure* used in the Fiction corpus (the 2nd GBN version)

After that, the authors of [10, 11] continued to use the GBN corpus in their studies of the language evolution [18, 19].

5   The Use of GBN

Despite the problems of the GBN corpus described above, it is widely used in various linguistic and culturological studies. There are over 6.500 articles mentioning GBN in the Google Scholar system. Approximately 200 works were published within the first 3.5 months of 2019. The review of those works is beyond the scope of this paper. However, we would like to mention some of the most interesting works which reflect modern trends in research and demonstrate ample opportunities of using GBN in the field of Digital Humanities.

The GBN data are used in different scientific fields. However, most studies are performed in linguistic domain. Language lexicon and its change are of primary interest to researchers. A variety of quantitative studies were performed to trace language dynamics and evolution. Among the investigated problems are the amount of words in the language vocabulary and its changes [20], dynamics of the "births" and "deaths" of words [21], the rate of evolution of different languages and their lexicon [16, 19]. Several studies were also conducted to reveal mechanisms of competing of regular and irregular forms of verbs in English [7] and compare British English and American English [16].

Regularities of language functioning were found using the GBN data. For example, [22] describes the way and rate of change of the core language vocabulary, and modelling of growth of syntactic relations network in English and Russian is performed in [23].

Semantic studies are also possible using the GBN corpora data. Methods of semantic change detection are introduced in [24–27]. Neural word embeddings were used in [24] for the first time to determine the word semantics and reveal semantic changes. They used the Skip-gram model trained by 5-grams of the GBN English fiction corpus. The methods described in [25, 26] also use neural word embeddings of different types and the GBN data. The method proposed in [27] is less complicated and uses data on frequencies of syntactic bigrams.

Besides linguistics, the GBN corpus is also used in psychology to study emotions [28–31] and cognitive processes [32–34]. Changes in psychology of collectivism/individualism turned out to be a widely discussed question. In our opinion, the most interesting articles in this area are [35–38] which investigate the growth of individualism in different countries using linguistic data from English, German, Russian, and Chinese.

The GBN corpus was also used in socio-cultural research. Studies of gender differences and diversity were conducted in [39, 40], and cultural trends were revealed in [41].

6   Directions in Enhancing the Results

The GBN corpus is an extremely large collection of texts and its improvement requires a lot of efforts. The first version was corrected and the second one was released in 2012. No one knows whether it is going to be further improved. However, there are several ways to make the result obtained using the GBN corpus more reliable.

The first one is to use all possible support data extracted from the corpus. For example, consider not just individual words but to investigate the behavior of its various inflectional forms [42] and even synonyms [37]. Such approach is demonstrated in [42] where the use of the word *eigen* is studied. It is shown that the form *eigen* (own, peculiar) is rarely found in the texts. However, frequency of the form *eigenen* is 35 times as much as frequency of the form *eigen*. Some scientists recommend studying each word along with its three synonyms selected from the relevant dictionaries of synonyms [37].

Sometimes it is relevant to perform comparative studies and see how the same or close in meaning terms are used in different corpora presented in GBN [42]. The English sub-corpus of GBN is divided into several subcorpora: general English, American English, British English and Fiction. They can be used to compare and verify the obtained results. For example, dynamics of the first-person pronoun frequencies is traced using both the American English and Fiction corpora [43].

The second way to enhance the results is to preprocess the GBN raw data though it is time-consuming and takes much efforts. For example, the corpus preprocessing that consists in removing all tokens (character strings) that are not words is described in [16]. All tokens that contain numbers or other non-alphabetic symbols, except for apostrophes, are deleted (the '–' symbol is processed by the GBN system itself).

GBN data preprocessing is especially useful for the languages that have undergone spelling reforms, such as the Russian language. The letter ъ was no more used after the Russian spelling reform of 1918. To process the words having the ending ъ correctly, it would be reasonable to delete the ending ъ from all the required words. This allows more reliable processing of an enormous number of Russian words (practically masculine nouns).

Other systematic changes in spelling can be revealed using the corpus and corrected in compliance with the current spelling rules, if necessary. Such approach is adopted in the RNC corpus where ancient orthography can be replaced by the modern one.

The GBN corpus is extremely large and it is impossible to correct all its errors. Therefore, the recently developed methods of working with the noisy language data can be used for the GBN data preprocessing [44, 45].

7   Conclusion

Corpus representativeness is a key issue in any linguistic study because corpus serves as a basis for generalizations concerning a language as a whole [17]. There are some recommendations for compiling a representative corpus. However, corpus building is a sophisticated task and perfectionism should be avoided in this field since no one knows what an ideal corpus would be like [46]. Anyway, corpora are a unique source of linguistic evidence and are useful for studying patterns and trends.

This article shows that the GBN corpus can be regarded as a representative one for the following reasons. It is the largest corpus ever existed which includes texts of various types and genres written by people of different age, sex and with different background. Such diversity of texts, their length and size serve a solid empirical foundation for linguistic and related studies. The GBN corpus is a diachronically balanced one since each year of the last decades is represented by a great number of texts ensuring wide range of linguistic distributions. The GBN corpus is not proportional. However, proportionality is less relevant for general linguistic studies than the range of texts presented in a corpus [17].

A comparative study of the results obtained using the GBN corpus and the representative corpora RNC and GICR showed that the GBN corpus can be successfully used for corpus-based studies. It was also shown that scientists may face some difficulties while working with the GBN corpus. Some of the problems were solved, such as metadata errors. Some of the problems still exist, for example ORC errors. However, great work has been performed to correct the ORC errors and now they cannot be viewed as a serious obstacle for research studies. Besides, the article describes how to make work with the GBN data more effective.

To sum up, we should say that the GBN corpus is not ideal as all text corpora ever compiled. However, it was shown that it can serve a good empirical foundation for certain linguistic and related studies. As it was said "it is better to be approximately right, than to be precisely wrong".

Acknowledgment. This research was financially supported by the Russian Foundation for Basic Research (Grant No. 17-29-09163), the Government Program of Competitive Development of Kazan Federal University, and through the State Assignment in the Area of Scientific Activities for Kazan Federal University, agreement № 34.5517.2017/6.7.

Appendix. The List of 100 Most Frequent Words with the ORC Errors

The original correct word form is given for each word in the brackets:

библногр (библиогр), сбориик (сборник), армш (армии), отношенш (отношении), производства (производства), иазв (назв), академш (академии), техиика (техника), населеш (населения), исследоваие (исследование), техиология (технология), проивводство (производство), траиспорт (транспорт), отношеш (отношения), основанш (основании), револющш (революции), оргаиизации (организации), стронтельство (строительство), великш (великий), состоянш (состоянии), историн (истории), стронт (строит), германш (германии), матерналы (материалы), движеш (движения), русскш (русский), англш (англии), оцеика (оценка), управлеш (управления), комиссш (комиссии), франщш (франции), траиспорта (транспорта), иекоторых (некоторых), положенш (положении), внимаш (внимания), работииков (работников), теорш (теории), управлеиия (управления), бнблиогр (библиогр), значеш (значения), выражеше (выражение), исследоваия (исследования), информацин (информации), издаше (издание), оргаиизациФ(организация), собраш (собрания), промышлениости (промышленности), явлеше (явление), револющи (революции), образоваия (образования), мииистров (министров), техиики (техники), существоваше (существование), организащш (организации), настроеше (настроение), оцеики (оценки), иауч (науч), зреш (зрения), экоиомика (экономика), географин (географии), сознаш (сознания), эффективиости (эффективности), уровия (уровня), коиструкций (конструкций), управлеше (управление), основаш (основания), территорш (территории), поэзш (поэзии), всякш (всякий), образоваш (образования), состояше (состояние), болезии (болезни), предложеше (предложение), информацни (информации), цеитр (центр), влияиие (влияние), студеитов (студентов), заключеше (заключение), отношешю (отношению), являшя (явления), италш (италии), издашя (издания), впоследствш (впоследствии), оргаиизаций (организаций), представлеше (представление), существовашя (существования), републжи (республики), бнол (биол), совершеиствование (совершенствование), примеиеиие (применение), теорни (теории), развитш (развитии), направленш (направлении), полезиых (полезных), деятельиости (деятельности), спасешя (спасения), промышлениость (промышленность), средией (средней), направлеше (направление), канцелярш (канцелярии).

References

1. Rykov, V.V.: Text corpus design as application of object-oriented paradigm. In: Trudy Mezhdunarodnogo seminara Dialog-2002, Nauka, Moskow, pp. 124–129 (2002). (in Russian)
2. Russian National Corpus. http://www.ruscorpora.ru. Accessed 29 Dec 2019
3. Natsional'nyy korpus russkogo yazyka: 2003–2005. Rezul'taty i perspektivy. Indrik, Moscow (2005). (in Russian)
4. Natsional'nyy korpus russkogo yazyka: 2006–2008. Novyye rezul'taty i perspektiv. Nestor-Istoriya, St. Petersburg (2009). (in Russian)
5. Belikov, V., Kopylov, N., Piperski, A., Selegey, V., Sharoff, S.: Corpus as language: from scalability to register variation. In: Computational Linguistics and Intellectual Technologies. Papers from the Annual International Conference "Dialogue", vol. 12, no. 19, pp. 83–95. RGGU, Moskow (2013)
6. Lin, Y., Michel, J.-B., Aiden, E.L., Orwant, J., Brockman, W., Petrov, S.: Syntactic annotations for the Google Books ngram corpus. In: 50th Annual Meeting of the Association for Computational Linguistics 2012, Proceedings of the Conference, vol. 2, pp. 169–174. Association for Computational Linguistics, Jeju Island, Korea (2012)
7. Michel, J.-B., Shen, Y.K., Aiden, A.P., Veres, A., Gray, M.K., et al.: Quantitative analysis of culture using millions of digitized books. Science 331(6014), 176–182 (2011)
8. Aiden, E., Michel, J.-B.: Uncharted Big Data as a Lens on Human Culture, 1st edn. Riverhead Books, New York (2013)
9. Belikov, V.I.: What and how can a linguist get from digitized texts? Siberian J. Philol. 3, 17–34 (2016). (In Russian)
10. Koplenig, A.: The impact of lacking metadata for the measurement of cultural and linguistic change using the Google Ngram data sets—Reconstructing the composition of the German corpus in times of WWII. Digit. Scholar. Human. 32, 169–188 (2017). https://doi.org/10.1093/llc/fqv037
11. Pechenick, E.A., Danforth, C., Dodds, P., Barrat, A.: Characterizing the google books corpus: strong limits to inferences of socio-cultural and linguistic evolution. PLoS ONE 10(10), e0137041 (2015)
12. Solovyev, V., Akhtyamova, S.: Linguistic big data: problem of purity and representativeness. In: XXI International Conference on Data Analytics and Management in Data Intensive Domains (DAMDID/RCDL 2019), Kazan, Russia, 15–18 October 2019, pp. 193–204. CEUR-WS.org (2019)
13. Dictionary OpenCorpora. http://opencorpora.org/dict.php. Accessed 29 Dec 2019
14. Bocharov, V.V., Alexeeva, S.V., Granovsky, D.V., Protopopova, E.V., Stepanova, M.E., Surikov, A.V.: Growdsourcing morphological annotation. In: Computational Linguistics and Intellectual Technologies. Papers from the Annual International Conference "Dialogue", vol. 12, no. 19, pp. 109–115. RGGU, Moskow (2013)

15. Solovyev, V.D.: Possible mechanisms of change in the cognitive structure of synonym sets. In: Language and Thought: In: Contemporary Cognitive Linguistics. A Collection of Articles, pp. 478–487. Languages of Slavic Culture, Moscow (2015). (in Russian)

16. Bochkarev, V., Solovyev, V., Wichmann, S.: Universals versus historical contingencies in lexical evolution. J. R. Soc. Interface **11**(101), 20140841 (2014). https://doi.org/10.1098/rsif.2014.0841

17. Biber, D.: Representativeness in corpus design. In: Zampolli, A., Calzolari, N., Palmer, M. (eds.) Current Issues in Computational Linguistics: In: Honour of Don Walker. Linguistica Computazionale, vol. 9, pp. 377–407. Springer, Dordrecht (1994). https://doi.org/10.1007/978-0-585-35958-8_20

18. Koplenig, A.: A fully data-driven method to identify (correlated) changes in diachronic corpora. arXiv preprint arXiv:1508/1508.06374 (2015)

19. Pechenick, E.A., Danforth, C., Dodds, P.: Is language evolution grinding to a halt? The scaling of lexical turbulence in English fiction suggests it is not. J. Comput. Science **21**, 24–37 (2017)

20. Petersen, A.M., Tenenbaum, J., Havlin, S., Stanley, H.E., Perc, M.: Languages cool as they expand: allometric scaling and the decreasing need for new words. Sci. Rep. **2**, 943 (2012)

21. Petersen, A.M., Tenenbaum, J., Havlin, S., Stanley, H.E.: Statistical laws governing fluctuations in word use from word birth to word death. Sci. Rep. **2**, 313 (2012)

22. Solovyev, V.D., Bochkarev, V.V., Shevlyakova, A.V.: Dynamics of core of language vocabulary. CEUR Workshop Proc. **1886**, 122–129 (2016)

23. Bochkarev, V.V., Shevlyakova, A.V., Lerner, E.Yu.: Modelling of growth of syntactic relations network in English and Russian. J. Phys.: Conf. Ser. **1141**, 012008 (2018). https://doi.org/10.1088/1742-6596/1141/1/012008

24. Kim, Y., Chiu, Y.-I., Hanaki, K., Hegde, D., Petrov, S.: Temporal analysis of language through neural language models. In: Proceedings of the 52nd Annual Meeting of the Association for Computational Linguistics, pp. 61–65. ACL, Baltimore (2014)

25. Kulkarni, V., Al-Rfou, R., Perozzi, B., Skiena, S.: Statistically significant detection of linguistic change. In: Proceedings of the 24th International Conference on World Wide Web, Florence, Italy, pp. 625–635 (2015)

26. Dubossarsky, H., Tsvetkov, Y., Dyer, C., Grossman, E.: A bottom up approach to category mapping and meaning change. In: Proceedings of the NetWordS Final Conference, Pisa, 30 March–1 April 2015, pp. 66–70. CEUR-WS.org (2015)

27. Bochkarev, V., Shevlyakova, A., Solovyev, V.: A method of semantic change detection using diachronic corpora data. In: van der Aalst, W.M.P., et al. (eds.) AIST 2019. CCIS, vol. 1086, pp. 94–106. Springer, Cham (2020). https://doi.org/10.1007/978-3-030-39575-9_10

28. Acerbi, A., Lampos, V., Garnett, P., Bentley, R.A.: The expression of emotions in 20th century books. PLoS ONE **8**(3), e59030 (2013). https://doi.org/10.1371/journal.pone.0059030

29. Mohammad, S.M.: From once upon a time to happily ever after: tracking emotions in mail and books. Decis. Support Syst. **53**(4), 730–741 (2012)

30. Morin, O., Acerbi, A.: Birth of the cool: a two-centuries decline in emotional expression in Anglophone fiction. Cogn. Emot. **31**(8), 1663–1675 (2017). https://doi.org/10.1080/02699931.2016.1260528

31. Scheff, T.: Toward defining basic emotions. Qual. Inq. **21**(2), 111–121 (2015)

32. Ellis, D.A., Wiseman, R., Jenkins, R.: Mental representations of weekdays. PloS ONE **10**(8), e0134555 (2015). https://doi.org/10.1371/journal.pone.0134555

33. Hills, T.T., Adelman, J.S.: Recent evolution of learnability in American English from 1800 to 2000. Cognition **143**, 87–92 (2015). https://doi.org/10.1016/j.cognition.2015.06.009

34. Virues-Ortega, J., Pear, J.J.: A history of "behavior" and "mind": use of behavioral and cognitive terms in the 20th century. Psychol. Rec. **65**(1), 23–30 (2015). https://doi.org/10.1007/s40732-014-0079-y

35. Greenfield, P.M.: The changing psychology of culture from 1800 through 2000. Psychol. Sci. **24**(9), 1722–1731 (2013). https://doi.org/10.1177/0956797613479387

36. Zeng, R., Greenfield, P.M.: Cultural evolution over the last 40 years in China: using the Google Ngram viewer to study implications of social and political change for cultural values. Int. J. Psychol. **50**(1), 47–55 (2015). https://doi.org/10.1002/ijop.12125

37. Younes, N., Reips, U.-D.: The changing psychology of culture in German-speaking countries: a Google Ngram study. Int. J. Psychol. **53**, 53–62 (2018). https://doi.org/10.1002/ijop.12428

38. Velichkovsky, B.B., Solovyev, V.D., Bochkarev, V.V., Ishkineeva, F.F.: Transition to market economy promotes individualistic values: analysing changes in frequencies of Russian words from 1980 to 2008. Int. J. Psychol. **54**, 23–32 (2019). https://doi.org/10.1002/ijop.12411

39. Del Giudice, M.: The twentieth century reversal of pink-blue gender coding: a scientific urban legend? Arch. Sex. Behav. **41**(6), 1321–1323 (2012). https://doi.org/10.1007/s10508-012-0002-z

40. Ye, S., Cai, S., Chen, C., Wan, Q., Qian, X.: How have males and females been described over the past two centuries? An analysis of Big-Five personality-related adjectives in the Google English Books. J. Res. Pers. **76**, 6–16 (2018)

41. Grossman, I., Varnum, M.: Social structure, infectious diseases, disasters, secularism, and cultural change in America. Psychol. Sci. **26**, 311–324 (2015)

42. Younes, N., Reips, U.-D.: Guideline for improving the reliability of Google Ngram studies: evidence from religious terms. PLoS ONE **14**(3), e0213554 (2019). https://doi.org/10.1371/journal.pone.0213554

43. Twenge, J.M., Campbell, W.K., Gentile, B.: Changes in pronoun use in American books and the rise of individualism, 1960–2008. J. Cross Cult. Psychol. **44**(3), 406–415 (2013)

44. Malykh, V., Lyalin, V.: Named entity recognition in noisy domains. In: Proceedings-2018 International Conference on Artificial Intelligence: Applications and Innovations, IC-AIAI 2018, vol. 8674438, pp. 60–65. IEEE (2018)

45. Malykh, V., Khakhulin, T.: Noise robustness in aspect extraction task. In: Proceedings-2018 International Conference on Artificial Intelligence: Applications and Innovations, IC-AIAI 2018, vol. 8674450, pp. 66–69. IEEE (2018)

46. Sinclair, J.: How to build a corpus. In: Wynne, M. (ed.) Developing Linguistic Corpora: A Guide to Good Practice, pp. 95–101. Oxbow Books, Oxford (2005)

Method for Expert Search Using Topical Similarity of Documents

Denis Zubarev[1]([✉]), Dmitry Devyatkin[1], Ilya Sochenkov[1,2], Ilya Tikhomirov[1], and Oleg Grigoriev[1]

[1] Federal Research Center "Computer Science and Control" of Russian Academy of Sciences, Moscow, Russia
zubarev@isa.ru

[2] Moscow State University, Moscow, Russia

Abstract. The article describes the problem of finding and selecting experts for reviewing grant applications, proposals and scientific papers. The main shortcomings of the methods that are currently used to solve this problem were analyzed. These shortcomings can be eliminated by analyzing large collections of sci-tech documents, the authors of which are potential experts on various topics. The article describes a method that forms a ranked list of experts for a given document using a search for documents that are similar in topic. To evaluate the proposed method, we used a collection of grant applications from a science foundation. The proposed method is compared with the method based on topic modeling. Experimental studies show that in terms of such metrics as recall, MAP and NDCG, the proposed method is slightly better. In conclusion, the current limitations of the proposed method are discussed.

Keywords: Scientific expertise · Expert search · Unstructured data analysis · Text analysis · Similar document retrieval

1 Introduction

A competent and objective examination of grant applications, research proposals and scientific papers is the important task of modern R&D. But it requires a competent and objective selection of experts. Currently, in most cases, the appointment of experts is based on manually assigned codes from manually created classifiers or manually chosen keywords. Experts and authors independently assign codes or keywords to their profiles or documents (application for a grant, report on a grant, an article, etc.), and the appointment of an expert is carried out by comparing the assigned codes or keywords. Classifiers are rarely updated, so they quickly become obsolete, have uneven coverage

The research is supported by Russian Foundation for Basic Research (grant №18-29-03087) The reported research is also partially funded by the project "Text mining tools for big data" as a part of the program supporting Technical Leadership Centers of the National Technological Initiative "Center for Big Data Storage and Processing" at the Moscow State University (Agreement with Fund supporting the NTI-projects No. 13/1251/2018 11.12.2018).

© Springer Nature Switzerland AG 2020
A. Elizarov et al. (Eds.): DAMDID/RCDL 2019, CCIS 1223, pp. 163–180, 2020.
https://doi.org/10.1007/978-3-030-51913-1_11

of the subject area (one code can correspond to thousands of objects, and the other to dozens) and have all the other drawbacks of manual taxonomies. In addition, experts often assign themselves several codes, but their level of competence varies greatly between these codes [1]. If there are several dozen experts who correspond to the same code (which happens quite often), then the further choice will be extremely subjective and non-transparent (in fact, manual selection of an expert is performed). All this leads to insufficient compliance of the competence of the selected expert and the object of examination and, possibly, to the subjective choice of the expert. As a result, there are refusals of examination, or it is conducted incompetently and, possibly, subjectively. Therefore, it is important not to determine the formal coincidence of the expert interests and the expertise subject topic, but to use all possible information for accurate expert ranking.

Information about expert competence is accumulated in the documents in which he/she participated (scientific articles, scientific and technical reports, patents, etc.). This information is much more precise in determining the expert's knowledge area than the classifier codes or keywords. This article describes the two methods of searching and ranking experts for a given object of expertise using thematically similar documents retrieval and using topic modeling. The methods require a database of experts and a large set of texts associated with experts. It is assumed that one of this method can become the basis for a whole class of methods that use unstructured information to select experts for the objects of expertise. The methods are compared in experimental studies using various metrics.

This paper is an extended version of our report [2] presented on the DAMDID 2019 conference. We added a comparison with a method based on topic modeling: we described how topic models can be used in the task of expert search and what hyperparameters of those models are significant for this task. We continued to improve an evaluation methodology: we introduced two new metrics that utilize available meta-information about experts and objects of expertise. In the end, we provided experimental comparison of two different topic models and the proposed method.

2   Related Works

Automating the search for an appropriate expert for examination has long been a subject of research. Researchers often narrow the research scope, for example, limiting themselves only to the appointment of experts to review articles submitted to the conference [3], or to select an expert who will answer user questions on the corporate knowledge base [4].

As a rule, expert assignment methods are divided into two groups [5]. The first group includes methods that require additional actions from experts or authors. For example, one of the methods involves the examination of the submitted abstracts of articles and the self-assessment of his readiness to consider any of the works in question. Another involves the selection by an expert of keywords that describe his competence from the list provided by the conference organizers and comparing these keywords from the expert with the keywords chosen by the authors of the article. These approaches are well-suited for small conferences but are not suitable for events in which several tens of thousands of participants take part. Even with relatively small conferences, the use of keywords is inappropriate if the number of topics for this event is large enough.

The second group includes methods that automatically build an expert's competence model based on his articles and/or other data and compare the resulting model with peer-reviewed articles submitted using the same model [6]. In this work, the name and surname of the expert were sent as a request to Google Scholar and CiteSeer. For the full texts of the articles found and the article submitted, the Euclidean distance was measured. This method does not take into account the namesakes, the dynamics of changes in the expert interests, a possible conflict of interests and requires significant computational resources. Another method [7] uses annotations and titles to classify articles according to topics predetermined by the conference organizers. However, it is not always possible to pre-determine a specific set of topics. The method presented in [8] uses bibliographic data from the reference list of the presented article. First names and surnames of authors are mined from bibliographic references, co-authors are determined for them using external resources (DBLP), etc. Thus, a co-authorship graph is constructed, on which a modification of the page ranking algorithm for identifying experts is performed. In [9] a special similarity measure that compares the reference lists are used to determine the proximity between expert publications and article submission. The comparison takes place under headings and authors, it also takes into account the case when the expert's articles are cited in the presented article. Bibliographic list comparison is a fairly effective operation, but it is difficult to assess the expert's competence only by bibliographic references, without using full texts. In [10] topic modelling is used to represent an object of examination and each document associated with an expert. Topic distribution of an expert is adjusted according to the time factor that is meant to capture the changes of research directions of an expert as time goes on. Cosine measure is used to measure the similarity between the expert's topic distribution and the topic distribution of the object under review. Furthermore, vector space model (with TF-IDF weighting scheme) is used to calculate an additional similarity score between experts and the object of examination. The final score for relevance is calculated using a weighted sum that takes into account the two previous scores. In the experimental studies of this work, the number of topics was chosen to be 100, which, according to the authors, reflects the real number of topics in information technology knowledge area. In this study, words are used as features.

In [11] a hybrid approach is used that combines full-text search (performed using ElasticSearch) over experts' articles and an expert profiling technique, which models experts' competence in the form of a weighted graph drawn from Wikipedia. The vertices of the graph are the concepts extracted from the expert's publications with TagMe tool. Edges represent the semantic relatedness between these concepts computed via textual and graph-based relatedness functions. After that, each vertex is assigned a score corresponding to the competence of the expert. This score is computed employing a random walk method. Concepts with a low score are removed from the expert's profile, due to the assumption that they cannot be used to characterize his competence. Also, the vertices are assigned a vector representation which is learned via structural embeddings techniques on concepts graph. At query time, the object of examination is parsed with TagMe tool, and embeddings are retrieved for extracted concepts, then they are averaged. As a result, a cosine measure is used to measure the similarity between averaged expert's vectors and vector that represents the object of the expertise. The final list of relevant experts is obtained via combining full-text search results and results of semantic profiles matching. It should be noted that impact of semantic profiles is rather small. According to the results of experimental studies conducted in this work, the increase in the quality assessment using the expert's semantic profile was 0.02, compared to the use of full-text search with the BM25 ranking function. This method was tested on a dataset [12], in which short phrases describe areas of knowledge (GT5). These phrases were used as queries (objects of expertise).

Thus, the existing methods for the expert search do not use all available information related to this task. Some methods are limited to processing only bibliographic lists or annotations with titles while ignoring the full texts of articles. Others are based on full text analysis of articles but they use ordinary full-text retrieval tools that apply to simple keyword search and are not effective for thematically similar documents retrieval. In addition, it should be noted that some of the methods described are computationally expensive since they do not use efficient means of indexation, and when selecting experts for each new object, it is necessary to repeat complex computational operations. In the approach proposed in this article, the main part of computationally intensive operations is performed only once.

3   Expert Search Methods

3.1   Similar Document Retrieval

We will refer to this method as "SDR" throughout the paper. The first step of the proposed method is the search for topically similar documents for a given object of examination (application) [13]. Search is made on the collections of scientific and technical texts. These can be scientific papers, patents and other documents related to/authored by experts. The collections are pre-indexed. Before indexing the text undergoes a full linguistic analysis: morphological, syntactic and semantic [14, 15]. Indexes store additional features for each word (semantic roles, the syntactic links and so on) [16]. During indexing, several types of indexes are created, including an inverted index of words and phrases, which is used to search for thematically similar documents. Indexing is incremental; that is, after initial indexing, one can add new texts to the collection without re-indexing the entire collection [16].

When searching for thematically similar documents, the given document is represented as a vector, elements of which are TF-IDF weights of keywords and phrases. Phrases are extracted based on syntactic relations between words. This allows extracting phrases consisting of words that are not adjacent to each other but have a syntactic connection. For example, the phrases "images search" and "digital images" will be extracted for the fragment: "search for digital images". The degree of similarity is calculated between the vector of the original document and the documents vectors from the index. Some similarity measure is used to calculate the degree of similarity (we tried cosine and hamming distance). The main parameters of the search method for thematically similar documents are presented in Table 1.

Table 1. The main parameters for thematically similar documents search method

Description	Name
The percent of words and phrases in the source document that determine the similarity of documents	`TOP_PERCENT`
The maximum number of words and phrases that are used to determine document similarity	`MAX_WORDS_COUNT`
The minimum number of words and phrases that are used to determine the similarity of documents	`MIN_WORDS_COUNT`
Minimum TF-IDF weight of a word or phrase included in the top keywords of the document	`MIN_WEIGHT`
The minimum value of the similarity score	`MIN_SIM`
The maximum number of similar documents for the source document	`MAX_DOCS_COUNT`

Based on the list of thematically similar documents, a list of candidates for experts is compiled. This is a trivial operation since the documents relate to the expert: the expert is one of the authors, he reviewed this article/application, etc.

After that, if there is the necessary meta-information, the experts are excluded from the list of candidates according to various criteria. For example, if meta-information about belonging to organizations is available for a peer-reviewed document and an expert, some experts could be filtered out because of a conflict of interest. At present, this step depends on the available meta-information, and it is related to the type of reviewed document. The experimental implementation of the method used several filters that are appropriate for grant applications:

1. All experts who are involved as participants in the given application are excluded.
2. All experts working in the same organizations as the head of the given application are excluded.

After that, the relevance of each expert to the object of expertise is calculated. The calculation takes into account the similarity of the documents (S_{sim}), with which the expert is associated, to the reviewed document, as well as several additional measures. In case if the expert has multiple documents, their ratings of similarity are averaged out. The set of additional measures depends on the type of the reviewed document. In the implemented method, one simple measure (S_{sci}) was used: the equality of the knowledge area code assigned to the expert and to the document under review (0 when the codes are not equal and 1 otherwise). The overall relevance score of the expert is calculated using the following formula:

$$W_{sim} \cdot S_{sim} + W_{sci} \cdot S_{sci}$$

where S_{sim}, S_{sci} – values of the measures described above, as W_{sim}, W_{sci} – weights with the condition $W_{sim} + W_{sci} = 1$. S_{sci} criterion was useful in ranking experts who were heads of interdisciplinary projects. An interdisciplinary project can relate to several scientific areas, but the head is an expert in only one area, so he should be ranked lower than the experts who have the same area of knowledge. The score of the relevance of each expert may lie in the interval [0; 1]. After evaluation, experts are ranked in descending order of relevance.

3.2 Topic Modeling

An expert usually has an interest in multiple research topics. Topic modeling allows representing experts and objects of expertise (papers, applications to grant) via topic distribution to characterize these research topics. Formally, each expert $e \in E$ is associated with a vector $\theta_{te} \in \mathbb{R}^T$ of T-dimensional topic distribution. Each element θ_{te} is the probability (i.e., $p(t|e)$) of the expert on topic t ($\sum_t \theta_{te} = 1$). In this way, each expert can be mapped onto multiple related topics. In the meantime, for a given object of expertise d, we can also find a set of associated topics. There are multiple possible ways to search the experts given the topic distribution of an object of expertise. One way is to form a list of experts such that the list can maximally cover the associated topics of the given object of expertise. Another way is to find experts that are competent enough in most topics of that object of expertise. In other words, find experts with the similar topic distribution as for the query document (e.g. via cosine similarity). We chose the second approach since the goal of the expert search task is to provide a large ranked list of relevant experts. Whereas the first approach is well suited for the final assignment of few experts (e.g. 3) for the given object of expertise.

Topic models usually are built from the document-terms matrix, which consists of columns corresponding to the documents and rows that correspond to the terms (words and phrases found in each document). So the result of training is the document-topic matrix and the term-topic matrix. It is possible to learn the topic distribution of documents' authors during training using some topic modeling frameworks, but we will not consider these models in this study. The trained topic model can be used to infer the topic distribution of an object of expertise using the term-topic matrix. We need to consider how to infer the topic distribution of experts by using the topic distribution of associated with experts documents. A natural way is to average all topic distributions of documents associated with the expert. Then the probability of an expert e belonging to the topic t can be expressed as follows

$$p(t|e) = \frac{\sum_i^{|D_e|} p(t|d_i)}{|D_e|}$$

where D_e denotes the set of documents associated with the expert e and $p(t|d_i)$ denotes the probability that the document d_i belongs to topic t. It is possible to avoid inferring of topic distribution of experts. It requires to search for documents with similar topic distribution, then to group documents by linked experts and to sort experts by averaged similarity scores. However, this approach gives worse results in conducted experimental studies.

On a preprocessing stage, each text was split into tokens, tokens were lemmatized, and syntax tree was built for each sentence. We used AOT for morphological and syntactic analysis. We used words and phrases as terms for building topic model. We used syntactic phrases up to 3 words in length. Namely, noun phrases where the modificator has specific part-of-speech: adjective, noun or various specific forms of verbs (participle, short participle etc.). The vocabulary was filtered based on document frequency (number of documents that contain a given term) with the following parameters: maximum document frequency rate = 0.4, minimum document frequency = 10, maximum dictionary size = 100000.

For comparison sake, experts with an obvious conflict of interest were removed from the ranked lists obtained by this method, based on meta-information available (like for SDR method).

4 Description of the Experimental Setup

4.1 Dataset Description

As a result of cooperation with the Russian Foundation for Basic Research, it was possible to conduct a series of experiments on the applications accumulated by the Foundation in various competitions held from 2012 to 2014. The Fund provided an API for indexing the full texts of applications. The application text included:

- summary of the project;
- description of the fundamental scientific problem the project aims to solve;
- goals and objectives of the study;
- proposed methods;
- current state of research in this field of science;
- expected scientific results;
- other substantive sections that are required by the competition rules.

For each application, a meta-information containing the following fields was provided:

- document identifier;
- the identifier of principal investigator;
- identifier of the organization in which the head works;
- a list of the identifiers of participants (co-investigators);
- coded participants full names;
- publication year of the grant application;
- code of the field of knowledge which the application belongs to (Biology, Chemistry, etc.));
- classifier codes;
- keywords of the application.

There was also presented impersonal information about the experts who reviewed the applications:

- identifier of an expert;
- identifier of the organization with which the expert is affiliated;

- keywords of an expert;
- code of the main area of knowledge of an expert;
- applications which the expert is the head in (list of identifiers);
- applications which the expert is the participant in (list of identifiers);
- applications reviewed by the expert (list of identifiers);
- applications the expert refused to review (list of identifiers).

The size of the collection of applications was about 65 thousand documents. Information was also received about 3 thousand experts, where the share of experts who were the head (principal investigator) of at least one application was 78%. At first, it was supposed to use only the applications of experts, in which they were principal investigators. However, it turned out that the share of such documents was about 9% among all grant applications. Moreover, most of the experts were associated with only one grant project. To increase the number of documents associated with the experts, we took into account applications in which the expert participated as co-investigator. We also used an external collection of scientific papers, which mainly consisted of articles from mathnet.ru and cyberleninka.ru, to search for additional experts publications. First, we looked for documents that confirm the support of grants with the participation of the expert (the grant identifier is usually written in the acknowledgments section). This provided us with about 4,000 additional documents. In addition, we performed a search for similar works for each expert application. To filter documents that are similar, but not related to experts, we compared the full names of the authors of the article with the full names of the applicants. If at least one full name corresponded, then we considered that this document is associated with an expert. Usually, there are no full names of the authors of the article, there are only short names (last name with initials), then there should be at least two matches with the short names of the applicants. We received about 30,000 new documents related to experts, using the search for similar documents.

Since the names of authors of papers are not structured and are presented as text, we parsed names into their individual components. We will briefly describe the parsing method. First, given the input string that contains the name of the author, the type of pattern is identified. Multiple patterns are supported:

1. The Slavic pattern includes several variations:
   a. Last name[,] First name [Patronymic];
   b. Last name Initials;
   c. First name [Patronymic] Last name;
   d. Initials Last name;
2. The Western pattern consists of several variations:
   a. First name [Middle]… Last name
   b. Last name, First name [Middle]…
   c. First name Initial [Initial]… Last name
3. Spanish pattern similar to the western one, except that there may be two last names:
   a. First last name [Second last name], First name [Middle]
   b. First name [Middle]… First last name [Second last name]
4. Asian pattern:
   a. Last name First name [First name]…

This classification is necessary because the parser can match full names with several patterns, e.g. 1.1 and 2.1. As a training set, we used the names of public persons and the country of their citizenship obtained from Wikidata dump, and also added the names with countries obtained from Russian patents. We trained Fasttext classifier on that dataset and obtained 0.96 precision@1 on the test data. When a pattern is identified for the given input text, then all variations available for this pattern are tested. If there is only one matching option, the parsing is complete. If more than one option matches, for example, 1.1 and 1.3; we use the common first names dictionary to determine the right variation.

After performing these procedures, the share of experts with documents increased to 88%. In addition, the number of experts associated with only one document was significantly reduced, as can be seen in Table 2.

Table 2. Distribution of the number of documents associated with an expert including extra documents

Number of documents per expert	Number of experts	Number of documents per expert	Number of experts	Number of documents per expert	Number of experts	Number of documents per expert	Number of experts
1	115	11	50	21	24	31	9
2	101	12	45	22	18	32	14
3	108	13	39	23	26	33	13
4	94	14	39	24	24	34	9
5	83	15	34	25	12	35	7
6	76	16	34	26	19	36	6
7	79	17	36	27	17	37	11
8	64	18	31	28	14	38	3
9	66	19	28	29	16	39	9
10	55	20	19	30	9	40	11

4.2 Evaluation Methodology

To assess the proposed method, data from previous expert selections of applications for participation in the A-2013 competition was used (total of 10,000 applications, an average of 3 experts per application). For every application from this competition, a ranked list of experts (found experts) was compiled using the proposed method. Then this list of experts was compared with the list of experts assigned to the given application.

There are common metrics that are used for evaluation of the experts search methods [17, 18]. Some of these metrics are applicable only if expert assignment goes along with the expert search. Those metrics assess the uniformity of the expert load and the assignment of a certain number of experts to each object of expertise. Each expert search provides a pool of relevant experts for further assignment. Therefore, this task should be evaluated using other metrics. Classical information retrieval metrics are frequently used: MAP, NDCG@100. Using these metrics is justified if the test data contains a large number of relevant experts for each object of expertise. We used a data set of up to 3 relevant experts for the object of expertise. This number of experts is not enough to correctly interpret the assessment results for a large number of selected experts. (like 100). Therefore, recall was used for evaluation in order to determine what total share of relevant experts was in the pool of selected experts. Recall was calculated using the following formula:

$$Recall = \frac{F_{found}}{F_{total}},$$

where F_{found} – the number of found experts from among those that have been assigned to this application; F_{total} – the number of assigned experts that could be found by the method (i.e. only experts that have at least one associated document).

Micro averaging was used to calculate metrics for all applications (i.e. for all applications are summarized F_{found} and F_{total}, and based on this, the required metric was calculated). Also, recall was calculated separately for each knowledge area.

The standard way of measuring precision in this situation is not appropriate since it is not known whether the found expert that has not been assigned to this application is suitable. He might be suitable for this application but was not assigned because he was busy on other projects or for other reasons.

Therefore, to evaluate precision, the information on this application expertise refusals was used. There were about 2 thousand of refusals according to the provided data. The idea is, that the compiled experts list shouldn't contain those who refused to review this application. The precision was calculated using the following formula:

$$Precision = 1 - \frac{R_{found}}{R_{total}},$$

where R_{found} – the number of found experts from those who refused to expertise the application; R_{total} – the number of refused experts that could be found by the method (i.e. experts with documents).

Since we do not have information about the relevance of the most retrieved experts, we decided to use extra meta-information of our dataset to evaluate relevance of found experts. We took into consideration the knowledge area, to which belongs an object of expertise and found expert: in most cases they should be the same.

The precision of retrieved knowledge areas for the given query document q is calculated by the formula:

$$P_{ar} = \frac{|\{e : e \in R \wedge Ar(e) = Ar(q)\}|}{|R|}$$

where R is a set of retrieved experts, and $Ar(e)$ is a function that returns assigned knowledge area code for the given object (expert or document). There are only 8 knowledge areas in our dataset. It is hardly possible to identify the subject of documents using those areas. Therefore, we also used codes of the RFBR classifier that are assigned to each document. It's possible to assign multiple codes to one document, but in our dataset one-third of documents have only one assigned code. There are about 360 unique codes in our dataset. To obtain codes of experts we made a union of all codes assigned to their applications. Then precision of the retrieved codes for the given query document q is calculated by the formula:

$$P_{cd} = \frac{|\{e : e \in R \wedge \Omega_e \cap Cd(q) \neq \emptyset\}|}{|R|}$$

where R is a set of retrieved experts, $Cd(d)$ is a function that returns a set of codes for a given document, and Ω_e is a set that is defined as follows: $\Omega_e = \bigcup\{Cd(d) : d \in D_e\}$ and D_e is a set of documents associated with expert e.

5 Parameters Tuning

5.1 SDR Method

Optimization of the algorithm parameters was performed on a separate sample collection of 700 applications. For optimization, we used a grid search of a single algorithm parameter with a fixed value of the remaining parameters. Optimization was performed to maximize recall. The results are presented in Table 3.

Table 3. Values of method parameters after optimization

Name	Value
TOP_PERCENT	0.4
MAX_WORDS_COUNT	200
MIN_WORDS_COUNT	15
MIN_WEIGHT	0.03
MIN_SIM	0.0
MAX_DOCS_COUNT	1000
W_{sci}	0.1
W_{sim}	0.9

We also tried various combinations of similarity functions and terms representation (only words, words with noun phrases); results are presented in Table 4.

Hamming distance along with adding word phrases result in the best recall. So these parameters were chosen.

5.2 Topic Modeling

Experiments were set up with two different models of topic modeling: Latent Dirichlet Allocation (LDA) [19] and Additive Regularization of Topic Models (ARTM) [20]. We

Table 4. Values of method parameters after optimization

	Recall
Cosine distance, only words	0.73
Cosine distance, with phrases	0.75
Hamming distance, only words	0.76
Hamming distance, with phrases	0.77

trained each model on a collection of documents associated with experts i.e. their papers and applications to grants.

For training ARTM model, we used BigARTM[1] library [21]. We selected the set of regularizers and hyperparameters of the model optimizing recall via grid search on the validation dataset. Optimized parameters are presented in the Table 5; the values in bold were chosen:

Table 5. Optimization results of ARTM model

Parameter	Value	Min recall	Max recall	Avg recall
Number of topics	50	0.698	0.713	0.705
Number of topics	100	0.728	0.740	0.734
Number of topics	**200**	0.734	0.760	0.744
Number of topics	300	0.730	0.763	0.742
Number of topics	400	0.716	0.763	0.735
Sparse phi regularizer	**−0.2**	0.707	0.763	0.737
Sparse phi regularizer	−0.1	0.704	0.757	0.732
Sparse phi regularizer	0.0	0.698	0.741	0.726
Sparse theta regularizer	−0.2	0.698	0.763	0.730
Sparse theta regularizer	**0.0**	0.698	0.763	0.731
Smooth theta regularizer	0.2	0.7	0.766	0.731
Decorrelator phi regularizer	15000	0.698	0.755	0.727
Decorrelator phi regularizer	1000	0.699	0.761	0.733
Decorrelator phi regularizer	**0**	0.700	0.763	0.734

It can be seen from the table that a small number of topics (50, 100) impacts negatively on model quality. The model learned for 200 topics shows the best average recall over all combinations of tuned parameters. Increasing the number of topics above 200 does not improve recall much. Sparsing all values in the term-topic matrix (sparse phi regularizer)

---

[1] https://github.com/bigartm/bigartm version 0.10.0.

works well for this task. It should be noted that adding this regularizer from the start of model learning worked better than adding it after some iterations. Sparsing (and smoothing) values in the topic-document matrix do not produce a significant impact on recall. The same is true for making topics more different with decorrelator phi regularizer.

For training LDA model we used Gensim[2] library [22]. We used the feature of Gensim library to learn parameters α and β from the corpus. Some hyperparameters were also optimized. They are presented in the table below (selected ones are in bold) (Table 6).

Table 6. Optimization results of LDA model

Parameter	Value	Min recall	Max recall	Avg recall
Number of topics	50	0.703	0.714	0.708
Number of topics	100	0.720	0.740	0.731
Number of topics	**200**	0.739	0.762	0.747
Number of topics	300	0.725	0.754	0.744
Number of topics	400	0.733	0.757	0.742
Document passes	**5**	0.703	0.762	0.743
Document passes	10	0.714	0.754	0.735
Document passes	20	0.706	0.739	0.725

It can be noted that the optimal number of topics is 200 as well as for the ARTM and 5 passes for each document is enough for learning topics for the task of expert searching.

6 Experimental Results

Table 7 shows the micro-averaged metrics (except MAP that is macro-averaged) on the top 170 of retrieved experts.

Table 7. Evaluation results

	Recall	Precision	MAP	NDCG	P_{ar}	P_{cd}
SDR	0.79	0.41	0.123	0.29	0.83	0.35
LDA	0.76	0.41	0.092	0.25	0.64	0.35
ARTM	0.76	0.42	0.093	0.25	0.63	0.35

All methods show similar performance in terms of recall and precision. The precision of retrieved knowledge areas (P_{ar}) is much better for the SDR method. It is due to

---

[2] https://github.com/rare-technologies/gensim version 3.8.1.

that SDR method ranks experts with the matching knowledge area higher than other experts. Interestingly that this does not have a substantial effect on recall and precision. The precision of retrieved classifier codes (P_{cd}) is expectedly lower than P_{ar}. As we discussed earlier, MAP and NDCG are not the best metrics for this task. It should be borne in mind that retrieved experts should be distributed over several dozens or hundreds of applications, and several experts are usually appointed for each application. Therefore, each application requires a sufficiently large pool of relevant experts. Therefore recall is a more important metric for this task than MAP or NDCG.

Since the result of the method is a ranked list of experts, it is possible to plot a graph of recall and precision shown in Fig. 1.

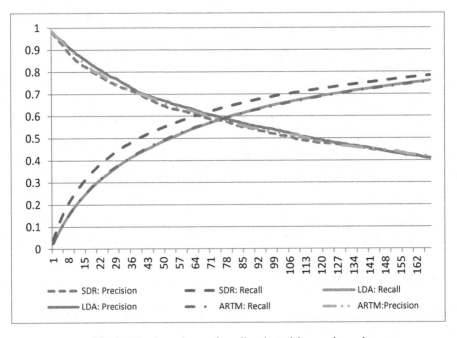

Fig. 1. The dependence of recall and precision on the rank

The graph shows a similar behaviour of all methods, with SDR being slightly better between 20 and 100 ranks. The maximum recall is achieved at 100-130 rank, after that, recall almost does not change.

Figures 2 and 3 show the dependencies on the rank of P_{ar} and P_{cd} respectively.

It is interesting that the precision of the topic models is slightly lower between 1 and 25 ranks in comparison with SDR. It can indicate that the most relevant experts with matching classifier codes are at the top of the ranked list. Whereas topic models produce a list where relevant experts are distributed evenly throughout the whole list. This better ranking of SDR method leads to better MAP and NDCG scores on test data.

Recall values were also calculated for each area of knowledge separately, and the results are shown in Table 8.

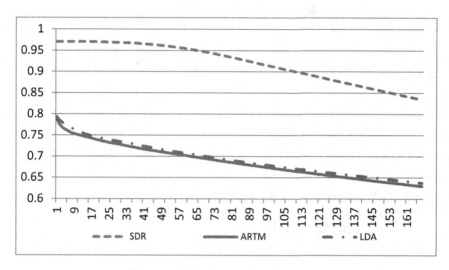

Fig. 2. The dependence of P_{ar} on the rank

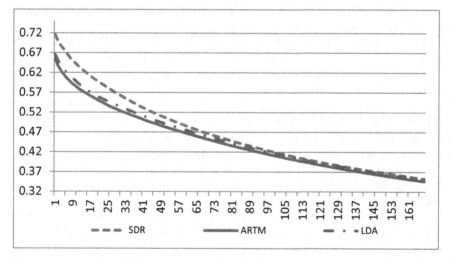

Fig. 3. The dependence of P_{cd} on the rank

Results differ in some areas. Therefore it can be beneficial to combine results obtained from two different methods. However, results are quite similar in terms of best-performing and worst-performing knowledge areas. The table shows that the best results were obtained in the fields of Earth Science (5) and the Human and Social Sciences (6) – recall is about 90%. In other areas, good results were obtained (recall from 70% to 80%). Average results were obtained for the fields of Information technologies (7) and Fundamentals of engineering (8).

Table 8. Recall for each knowledge area

	Number of applications	SDR	LDA	ARTM
Mathematics informatics and mechanics	969	0.78	0.75	0.72
Physics and astronomy	1462	0.82	0.78	0.78
Chemistry	1339	0.8	0.73	0.72
Biology and medical science	2183	0.78	0.79	0.81
Earth Sciences	1168	0.89	0.88	0.89
Human and Social Sciences	898	0.84	0.95	0.95
Information technologies and computer systems	1036	0.65	0.6	0.55
Fundamentals of Engineering	1448	0.68	0.61	0.64

7   Conclusion

In this paper, the expert search method based on the analysis of text information was described, and the results of method comparison with the method based on topic modeling were presented. We proposed a new evaluation methodology and conducted experiments on the RFBR data set, which distinguishes this work from the previous ones. The method showed better recall, MAP and NDCG than other methods. The further directions of research can be an extension of the method to support the inclusion of documents indirectly associated with the expert, for example, articles that he reviewed. These documents should contribute to the overall score of relevance with a lower ratio since the expert has no direct relationship to the text, however, if he regularly reviews the papers of a certain topic, it should be taken into account when scoring.

In further experiments, it is also proposed to expand the set of criteria that affect the expert's assessment for a given object of examination, for example, add an expert rating calculated using a page ranking algorithm based on quotes from expert works.

Further studies are also expected to improve the methodology for evaluation of the expert's assignment. In the technique proposed in the article, there are several shortcomings, namely: precision measurement based on refusals of expertise is not optimal. Cases of refusal are 15 times less than cases of acceptance, and refusal can occur for other reasons than a mismatch between the competence of the expert and the subject of the application. However, the question of how to evaluate the work of the expert search method is currently unresolved. The involvement of external experts can significantly improve the quality of the evaluation, but it will require a large number of experts from different knowledge areas, who should be well acquainted with the expert community.

The proposed method can be used not only when searching an expert for grant application of a scientific fund but also in the reviewer selection for any text object: articles in scientific journals, conference abstracts, patent applications, etc.

References

1. The Russian Scientific Foundation held a meeting of the expert council on scientific projects. http://rscf.ru/ru/node/2367. Accessed 02 Mar 2020
2. Zubarev, D.V., Devyatkin, D.A., Sochenkov, I.V., Tikhomirov, I.A., Grigoriev, O.G.: Expert assignment method based on similar document retrieval from large text collections. In: CEUR Workshop Proceedings of the Data Analytics and Management in Data Intensive Domains: Selected Papers of the XXI International Conference on Data Analytics and Management in Data Intensive Domains (DAMDID/RCDL 2019), vol. 2523, pp. 266–278 (2019)
3. Dumais, S.T., Jakob N.: Automating the assignment of submitted manuscripts to reviewers. In: Proceedings of the 15th Annual International ACM SIGIR Conference on Research and Development in Information Retrieval, pp. 233–244. ACM (1992)
4. Balog, K., Azzopardi, L., De Rijke, M.: Formal models for expert finding in enterprise corpora. In: Proceedings of the 29th Annual International ACM SIGIR Conference on Research and Development in Information Retrieval, pp. 43–50, ACM (2006)
5. Kalmukov, Y., Boris, R.: Comparative analysis of existing methods and algorithms for automatic assignment of reviewers to papers (2010). arXiv preprint https://arxiv.org/pdf/1012. 2019.pdf. Accessed 11 May 2019
6. Pesenhofer, A., Mayer, R., Rauber, A.: Improving scientific conferences by enhancing conference management systems with information mining capabilities. In: 2006 1st International Conference on Digital Information Management, pp. 359–366. IEEE (2006)
7. Ferilli, S., Di Mauro, N., Basile, T.M.A., Esposito, F., Biba, M.: Automatic topics identification for reviewer assignment. In: Ali, M., Dapoigny, R. (eds.) IEA/AIE 2006. LNCS (LNAI), vol. 4031, pp. 721–730. Springer, Heidelberg (2006). https://doi.org/10.1007/11779568_78
8. Rodriguez, M.A., Bollen, J.: An algorithm to determine peer-reviewers. In: Proceedings of the 17th ACM conference on Information and knowledge management, pp. 319–328. ACM (2008)
9. Li, X., Watanabe, T.: Automatic paper-to-reviewer assignment, based on the matching degree of the reviewers. Procedia Comput. Sci. **22**, 633–642 (2013)
10. Peng, H., Hu, H., Wang, K., Wang, X.: Time-aware and topic-based reviewer assignment. In: Bao, Z., Trajcevski, G., Chang, L., Hua, W. (eds.) DASFAA 2017. LNCS, vol. 10179, pp. 145–157. Springer, Cham (2017). https://doi.org/10.1007/978-3-319-55705-2_11
11. Cifariello, P., Ferragina, P., Ponza, M.: Wiser: a semantic approach for expert finding in academia based on entity linking. Inf. Syst. **82**, 1–16 (2019)
12. Berendsen, R., et al.: On the assessment of expertise profiles. J. Am. Soc. Inf. Sci. Technol. **64**(10), 2024–2044 (2013)
13. Sochenkov, I.V., Zubarev, D.V., Tihomirov, I.A.: Exploratory patent search. Inform. Appl. **12**(1), 89–94 (2018)
14. Osipov, G., et al. Relational-situational method for intelligent search and analysis of scientific publications. In: Proceedings of the Integrating IR Technologies for Professional Search Workshop, pp. 57–64 (2013)
15. Shelmanov, A.O., Smirnov, I.V.: Methods for semantic role labeling of Russian texts. In: Proceedings of International Conference Dialog on Computational Linguistics and Intellectual Technologies, vol. 13, no. 20, pp. 607–620 (2014)
16. Shvets, A., et al.: Detection of current research directions based on full-text clustering. In: 2015 Science and Information Conference (SAI), pp. 483–488. IEEE (2015)
17. Li, L., Wang, L., Zhang, Y.: A comprehensive survey of evaluation metrics in paper-reviewer assignment. In: Computer Science and Applications: Proceedings of the 2014 Asia-Pacific Conference on Computer Science and Applications (CSAC 2014), Shanghai, China, 27–28 December 2014, p. 281. CRC Press (2014)

18. Lin, S., Hong, W., Wang, D., Li, T.: A survey on expert finding techniques. J. Intell. Inf. Syst. **49**(2), 255–279 (2017). https://doi.org/10.1007/s10844-016-0440-5
19. Blei, D.M., Ng, A.Y., Jordan, M.I.: Latent Dirichlet allocation. J. Mach. Learn. Res. **3**(Jan), 993–1022 (2003)
20. Vorontsov, K., Anna, P.: Additive regularization of topic models. Mach. Learn. **101**(1–3), 303–323 (2015)
21. Vorontsov, K., Frei, O., Apishev, M., Romov, P., Dudarenko, M.: BigARTM: open source library for regularized multimodal topic modeling of large collections. In: Khachay, MYu., Konstantinova, N., Panchenko, A., Ignatov, D.I., Labunets, V.G. (eds.) AIST 2015. CCIS, vol. 542, pp. 370–381. Springer, Cham (2015). https://doi.org/10.1007/978-3-319-26123-2_36
22. Rehurek, R., Sojka, P.: Software framework for topic modelling with large corpora. In: Proceedings of the LREC 2010 Workshop on New Challenges for NLP Frameworks (2010)

Depression Detection from Social Media Profiles

Maxim Stankevich[1]([✉]), Ivan Smirnov[1,2], Natalia Kiselnikova[3],
and Anastasia Ushakova[4]

[1] Federal Research Center "Computer Science and Control" of RAS, Moscow, Russia
`{stankevich,ivs}@isa.ru`
[2] Peoples' Friendship University of Russia (RUDN University), Moscow, Russia
[3] Psychological Institute, Russian Academy of Education, Moscow, Russia
`nv.pirao@gmail.com`
[4] Moscow Institute of Physics and Technology, Moscow, Russia
`ushakova.av@phystech.edu`

Abstract. The problem of early depression detection is one of the most important in the field of psychology. Social network analysis is widely applied to address this problem. In this paper, we consider the task of automatic detection of depression signs from textual messages and profile information of Russian social network VKontakte users. We describe the preparation of users' profiles dataset and propose linguistic and profile information based features. We evaluate several machine learning methods and report experiments results. The best performance in our experiments achieved by the model that was trained on features that reflects information about users' subscriptions on Vkontakte groups and communities.

Keywords: Depression detection · Social networks · Psycholinguistics

1 Introduction

Nowadays the problem of early depression detection is one of the most important in the field of psychology. Over 350 million people worldwide suffer from depression, which is about 5% of the total population. Close to 800 000 people die due to suicide every year and it is statistically the second leading cause of death among people in 15–29 years old [1, 2]. At the same time, the major number of suicides associated with depression. Recent researches reveal that depression is also the main cause of disability and a variety of somatic diseases.

For example, Beliakov [3] in his paper summarizes the main results of recent depression, anxiety, and stress investigations and their relation to cardiovascular mortality. His overview shows that increased risk of death from cardio-vascular diseases associates with depression and stress. Surtees et al. conducted a prospective study in the UK that based on the 8.5 years of observation [4]. This study provides that the presence of major depression is associated with a 3.5-fold increase in mortality from coronary heart disease (CHD). Whang et al. demonstrate that women with depression have an increase of fatal CHD by 49% in 9 years of follow-up [5]. These studies demonstrate that depression

© Springer Nature Switzerland AG 2020
A. Elizarov et al. (Eds.): DAMDID/RCDL 2019, CCIS 1223, pp. 181–194, 2020.
https://doi.org/10.1007/978-3-030-51913-1_12

treatment and stress control, as well as early diagnosis and prevention of symptoms of psychological distress and mental disorders, can increase life expectancy.

Nevertheless, depression is still often falsely associated with a lack of will-power and unwillingness to cope with the "bad mood". There is a social stigmatization of this disease, and it is embarrassing to admit it for a person. As a result, people with depression often hide their condition, do not seek help in time, and aggravate the disease.

Online methods and social media provide an opportunity to privately detect the symptoms of depression in time. It would allow people to suggest measures for its prevention and treatment in the early stages. The report of the European branch of WHO (2016) paid special attention to the identification of signs of depression and the personalization of online methods of its prevention.

In our previous work, we considered the problem of automatic detection of depression signs from textual messages in Russian-speaking social network Vkontakte [6]. We retrieved psycholinguistic and stylistic features from users' texts and defined binary classification task using Beck Depression Inventory screenings. Several machine learning models were trained to evaluate the applicability of proposed features for the task.

This paper presents an extended and revised version of our previous work and we want to outline changes. First of all, the dataset for this research was extended with new samples because we continued to collect data from Vkontakte users since our previous work. We impose some changes on data preparation steps and retrieved additional features from users' data. The following text-based features sets were evaluated: psycholinguistic markers, POS-tags, syntax relations, semantic roles and relations, dictionaries, and n-grams. As a most notable change, we formed two features sets based on profile information of users. The first one contains the basic profile information (number of friends, number of followers, number of likes, etc.) and the second one is based on the information about users' subscriptions on different Vkontakte groups and communities. Since we collected more data volume, the design of experiments was also changed and now represented as a binary classification task with stationary train/test data split (in contrast to cross-validation results in previous work). The structure of works remained the same: in Sect. 2 related works are reviewed, in Sect. 3 we present dataset of Vkontakte profiles, in Sect. 4 we describe our methods and feature engineering and in last sections, we present and discuss results of experiments.

2   Related Work

Instrumental possibilities of analyzing the behavior of users in social networks are actively developing. In particular, methods of computational linguistics are successfully used in analyzing texts from social networks.

The computerized analysis method of texts LIWC (Linguistic Inquiry and Word Count) [7] allows assessing the extent to which the author of a text uses the words of psychologically significant categories. The method works on the basis of manually compiled dictionaries of words that fall into different categories: meaningful words (social, cognitive, positive/negative words, etc.), functional words (pronouns, articles, verb forms, etc.). LIWC is used for different languages, including Russian [8], but does not consider the specifics of the language, since it is simply a translation of dictionaries from English to Russian.

Yates et al. [9] used a neural network model to reveal the risks of self-harm and depression based on posts from Reddit and Twitter and showed the high accuracy of this diagnostic method. The authors indicate that proposed methods can be used for large-scale studies of mental health as well as for clinical treatment.

Seabrook et al. [10] utilized the MoodPrism application to collect data about status updates and the mental health of Facebook and Twitter users. It was found that the average proportion of words expressing positive and negative emotions, as well as their variability and instability of manifestation in the status of each user, can be used as a simple but sensitive measure for diagnosing depression in a social network. In addition, it was found that the use of the proposed method may depend on the platform: for Facebook users, these features predicted a greater severity of depression, and lower for Twitter.

Al-Mosaiwi et al. [11] examined the usage of absolute words (i.e., always, totally, entire) in text writings from various forums devoted to different disorders: depression, anxiety, suicidal ideation, posttraumatic stress disorder, eating disorder, etc. It was found that the number of absolute words in anxiety, depression, and suicidal ideation related forums was significantly greater than in forums from the control group.

Most of the related studies investigate the relationship between mental health and English-speaking social media texts. As an exception, Panicheva et al. [12] and Bogolyubova et al. [13] investigated the relationship between the so-called dark triad (Machiavellianism, narcissism, and psychopathy) and Russian texts from Facebook. Using the results of the dark triad questionnaire and profile data of Facebook users the authors conducted a correlation analysis to reveal informative morphological, lexical, and sentiment features.

The study of detecting an early risk of depression based on the experimental task Clef/eRisk 2017 described in the article [14]. The main idea of the task was to classify Reddit users into two groups: the case of depression and non-risk case. The study evaluates the applicability of tf-idf, embeddings, and bigrams models with stylometric and morphological features using the Clef/eRisk 2017 dataset and report 63% of F1-score for depression class.

It should be noted that the use of computational linguistics for analyzing text messages of social networks is mainly limited to lexical approaches. The syntactic-semantic analysis and psycholinguistics markers of the text are still not well evaluated on depression detection task. In this paper, we evaluated the various list of text based features as well as features that based on social media profile data.

3 Dataset

We asked volunteers from Vkontakte to take part in our psychological research and complete the Beck Depression Inventory questionnaire [15]. This questionnaire allows calculating depression scores on the 0–63 scale. Before answering questions, users gave access to their public pages under privacy constraints via the Vkontakte application. We automatically collected all available information from public personal profile pages using Vkontakte API for the users who completed the questionnaire. Posts, comments, information about communities, friends, etc. were collected from January 2017 to April

2019 for each user. Overall, information from 1330 profiles was assembled to compile our dataset. All of the personal information that can reveal the identity of persons were removed from data collection. General statistics on collected data presented in Table 1.

Table 1. General statistics on collected data.

Number of users	1330
Age	24.74 ± 6.84
Males	425 (32%)
Females	904 (68%)
Depression score	18.79 ± 11.56
Post count	94660

The scope of our interest were mainly textual messages, namely posts, written in Russian. Therefore, we focused on text messages written by Vkontakte users on their personal profiles and mainly operate with these messages. It is important to note, that social media data contains a significant amount of noise and text volume for each user considerably varies from person to person. Before performing on the depression detection task, we accurately cleaned the data. First, we applied constrains on the required text volume and number of posts. Secondly, we analyzed scores from Beck Depression Inventory and divided our users into 2 groups: persons with a score less than 11 were annotated as the control group (users without depression signs); persons with a score greater than 29 were annotated as depression group (users with depression signs). In this section, we describe these steps and provide statistics on the data. We refer to the data before any changes as initial data and to the data after cleaning and depression score grouping as classification data.

The initial data contained information about 1330 persons who took the Beck Depression Inventory questionnaire. The distribution of Beck Depression inventory scores across users from initial data presented in Fig. 1.

The mean age in the initial data is 25. The gender partition is unbalanced: 904 (68%) Females and 425 Males (32%). We already noted from our previous work, that initial data is extremely noisy. Standard deviation values for post, sentence, and word counts are doubled in comparison with their mean values. It was also discovered that more than 200 users from the dataset did not provide any textual volume. The superficial analysis of the data revealed that data require adjustments and cleaning. As the next step, we performed several actions to adjust the data:

1. Removed all characters which are not alphabet or standard punctuation symbols from texts using regular expressions;
2. Removed all posts with more than 4500 characters or less than 2 characters;
3. Removed all users with less than 500 characters provided;
4. Considered only the first 60000 characters per user.

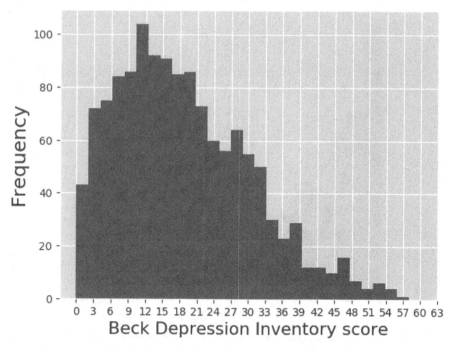

Fig. 1. Depression scores distribution in *initial data.*

In contrast to our previous work, we applied less limitation on the data. The manual analysis of posts with length more than 4500 characters revealed than long posts mainly not authored by users but contains text content from Internet resources. Applying these steps to the initial data yielded 816 user profiles. The post count distribution presented in Fig. 2.

After the data cleansing stage, we found this text volume much more suitable for applying natural language processing tools and performing any type of machine learning evaluation. Anyhow, the depression scores provided by Beck Depression Inventory required some interpretation. We outlined 2 different ways of how we can design our research. The first one is the regression analysis using raw depression scores, which might be seen as the most appropriate and confident way. But on the other hand, most of the English language based depression detection tasks were designed as a binary classification problem: discover if a person depressed or not. Moreover, binary classification results are easier to interpret. To make it possible to compare our results, we decided to perform a similar binary classification task on given data and compare our results with Clef/eRisk 2017 Shared Task [16].

As the next step, we analyzed depression scores and discovered that we cannot simply divide our data by setting border value and annotating all users with depression scores above this value as a risk group of depression and all users with the depression score bellow the border value as a non-risk group. In order to form the classification data, we annotated all persons with depression score less than 11 as a non-risk group (control group). For a risk group, we assembled the data of persons with depression scores above 29 (depression group). These values were discussed and proposed by the psychologist

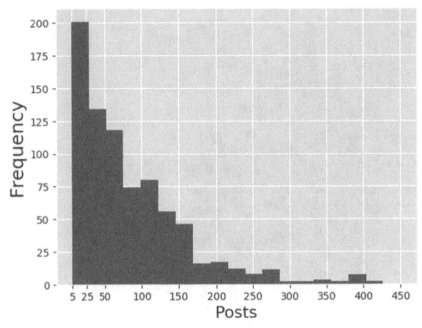

Fig. 2. Posts count distribution in *cleaned data*

experts related to our study. The persons with depression score between these values were removed from observation.

Performing this step reduced the data population to the 387 users, where 239 were labeled as the control group (without depression signs) and 148 users were labeled as belonging to the depression group. The statistics between groups on the classification data presented in Table 2.

Table 2. Statistics between depression and control group.

Group	Depression group	Control group
Number of users	148	239
Males	30 (20%)	83 (34.72%)
Females	118 (80%)	156 (65.27%)
Age	25.04 ± 6.76	25.97 ± 6.03
Depression score	36.29 ± 6.08	6.01 ± 2.81
Total number of posts	10693	20706
Avg. posts count	72.25 ± 81.74	86.74 ± 69.27
Avg. words count	1352.96 ± 2007.62	1972.56 ± 2481.88
Avg. sentences count	169.21 ± 212.08	218.40 ± 221.74
Avg. words per post	23.12 ± 25.80	23.86 ± 31.58
Avg. words per sentence	7.77 ± 4.16	7.91 ± 2.85
Avg. sentences per post	2.71 ± 2.62	2.71 ± 3.20

It can be observed from Table 2 that users from the depression group tend to write a lesser amount of text in the Vkontakte social media. The values of average posts count, average word count, and average sentence count are less than in the control groups. The gender partition is even more biased towards females in the depression group.

4 Features and Methods

4.1 Feature Sets

Before forming the feature sets, all user posts were concatenated into the one text for every user in the dataset. We applied MyStem [17] for tokenization, lemmatization, and part-of-speech tagging, and Udpipe [18] for syntax parsing. The sentiment features were calculated using the Linis-Crowd sentiment dictionary [19].

Psycholinguistic Markers
Psycholinguistic markers are linguistic features of a text that represent the psychological characteristics of the author and may signal about his psychological disorders. For example, people in stress more frequently use the pronoun "we" [20]. Psycholinguistic markers are calculated on morphological and syntactic information and in a manner corresponding to the writing style of the author. We use more than 30 markers and the most significant of them are the following:

- Mean number of words per sentence
- Mean number of characters per word
- (N punctuation characters)/(N words)
- Lexicon: (N unique words)/(N words)
- Average syntax tree depth
- (N verbs)/(N adjectives)
- (N conjunctions + N prepositions)/(N sentences)
- (N infinitives)/(N verbs)
- (N singular first person past tense verbs)/(N verbs)
- (N first person verbs)/(N verbs)
- (N third person verbs)/(N verbs)
- (N first person pronouns)/(N pronouns)
- (N singular first person pronouns)/(N pronouns)
- (N plural first person pronouns)/(N pronouns)

These psycholinguistic markers were previously utilized for the task of predicting depression from essays in Russian. They are described in more details at [21]. We extend psycholinguistic markers with features based on Linis-Crowd sentiment dictionaries as well as with following features that are specific for the social network: uppercase characters ratio, number of exclamation marks, number of "sad" and "happy" smiles.

Part-of-Speech Tags
As another feature, we utilized postags ratio in users' texts. For each user, we calculated the proportion of all parts of speech usage relative to the provided number of words in

the user's writings. We already applied such feature on the Clef/eRisk 2017 Shared Task data in [14].

Dictionaries

Another feature set was retrieved by utilizing dictionaries that were used for the task of detection verbal aggression in social media writings [22]. It is containing following dictionaries: negative emotional words, lexis of suffering, positive emotional words, absolute and intensifying terms, motivation and stressful words, invectives, etc. To calculate features, for every user we calculate the occurrences of words from different dictionaries in user's posts and divide this number on total user's words count.

Syntax Relations

This feature set was calculated by using syntax relations that were annotated in users' texts by Udpipe pipeline trained on SynTagRus treebank [23]. The total number of each relation was divided by the number of sentences and represented as a feature vector for each user.

Semantic Roles and Relations

In general words, a semantic role is an underlying relationship that a participant has with the main verb in a clause. To retrieve semantic roles and relations we applied the method described in [24]. The semantic features were calculated in a similar way as it was calculated for part of speech tags and syntax relations (relative to the text volume provided by each user in data).

N-Grams

We formed two n-grams sets: unigrams and bigrams. N-grams that appeared less than in 6 texts or more than in 80% of texts were removed from the feature sets. Overall, the lexicon contains 10067 and 5363 items for unigrams and bigrams respectively. We used tf-idf values formed on the basis of these n-gram models as 2 different feature sets.

Profile Information

In contrast to our previous version of this work where we evaluate only text features, we decided to observe features from profile information of users. We retrieved following profile features: number of friends, number of followers, number of groups and communities, number of photos/audios/videos, repost ratio (% of posts authored by user), likes per each post, gender, age, answers on standard predefined Vkontakte profile question (Attitude towards smoking and alcohol, relation status), and some binary features that reflects completeness of profile page. Some of these features were transformed using one-hot encoding.

Subscription Matrix

As another feature that is specific to the social media data, we assembled the information about users' subscriptions on different Vkontakte groups, communities, and popular profiles. First of all, we gathered the list of 20973 IDs from users' subscriptions. The groups, communities and popular profiles with overall subscription count less than 4 users from our data were removed from observation. This step reduced the length of the

IDs list to 1615 items. On the basis of this IDs list, for each user, we formed a feature vector with 1615 dimensions by filling the elements with binary 0 and 1 according to if observed user subscribed to the corresponding Vkontakte group, community or person.

4.2 Evaluation Setup

To evaluate the proposed features, we applied the train/test split with 75% for train samples and 25% for test samples. The train set consists of 173 control and 117 depressed users. The test set consists of 66 controls and 31 depressed. The short annotation for feature groups described in Table 3.

Table 3. Feature sets annotations.

Feature set	Annotation
Psycholinguistic markers	PM
Part-of-speech tags	POS
Dictionaries	D
Syntax relations	S
Semantic roles and relations	SRL
Unigrams	UG
Bigrams	BG
Profile information	PI
Subscriptions matrix	SM

5 Results of Experiments

To perform the classification, the estimators from the scikit-learn [25], xgboost [26] and lightgbm [27] packages were used. The following machine learning methods were evaluated:

- Support Vector Machine (SVM);
- Random Forest (RF);
- Logistic Regression (LR);
- Multilayer Perceptron (MLP)
- Naive Bayes (NB);
- Adaptive Boosting (AB);
- LightGBM Boosting (LGBM);
- XGBoost (XGB).

The feature's sets were normalized and scaled before used to train the models. Hyper-parameters of the classification algorithms were tuned by grid-search with 5-fold cross-validation on the train data. Since the depression detection task is previously untested on the Russian-speaking social media data, we also demonstrate the accuracy yielded by a random based dummy classifier (DC) with a stratified strategy. The metrics for evaluation are recall, precision, F1-score for depression group class and ROC AUC score. The main metric is the F1-score. All of the mentioned classifiers were tuned and trained on train data with different feature sets and all evaluations were completed on test data. In Table 4 we outlined best performances that were achieved on each set.

Table 4. Classification results different feature sets.

Feature set	Classifier	AUC	Precision	Recall	F1-score
–	DC	.48	.30	.29	.30
PM	XGB	.71	.52	**.74**	.61
POS	RF	.66	.50	.61	.55
D	SVM	.68	.49	.71	.58
S	MLP	.54	.36	.48	.41
SRL	RF	.62	.47	.52	.49
UG	RF	**.72**	**.53**	.74	**.62**
BG	RF	.65	.48	.65	.55
PI	NB	.62	.41	.74	.53
SM	NB	.65	.44	.74	.55

The best result in Table 4 yielded by RF with unigram feature set (.62 F1-score and .72 ROC AUC score). The second result achieved by PM set with SVM based classifier (.61 F1-score). The dictionaries feature set also achieved comparable results. The syntax relation feature set performed extremely poorly on the task.

Since SM features set is represented as a sparse matrix, we decided to apply dimensionality reduction on this set. We performed classification on SM set with PCA [28] transformation. The following number of PCA components were considered: 2, 5, 10, 20, 30, 40, 50, 100. The hyperparameters were tuned similarly to the first experiment, and the same metrics for the test set were calculated. In Table 5 we outline 10 best performances yielded by classifiers on test data.

The classification results in Table 5 revealed that applying PCA dimensionality reduction significantly improves classification performance. The best result was achieved by RF model with 10 PCA elements (.65 F1-score).

As a next step, we formed 3 different voting classifiers. Voting classifiers were utilized to wrap the models that were tuned on different sets of features and make predictions according to the majority rule. The sub-estimator combinations for voting classifiers were randomly chosen between models with the best 5-fold cross-validation

Table 5. Result of classification on SM feature set transformed by PCA for dimensionality reduction.

Classifier	AUC	Precision	Recall	F1-score	N components
RF	*.74*	*.59*	*.71*	*.65*	*10*
RF	*.72*	*.71*	*.55*	*.62*	*100*
AB	*.72*	*.57*	*.68*	*.62*	*5*
RF	*.70*	*.54*	*.68*	*.60*	*20*
LR	*.69*	*.51*	*.68*	*.58*	*50*
XGB	*.68*	*.48*	*.74*	*.58*	*20*
XGB	*.69*	*.53*	*.65*	*.58*	*100*
SVC	*.67*	*.44*	*.84*	*.58*	*10*
LR	*.69*	*.54*	*.61*	*.58*	*30*
SVM	*.67*	*.45*	*.81*	*.57*	*50*

F1-score. Another sub-voting classifier that was included in all voting classifiers was used to create an estimator with higher precision on the data. We report results by 3 different voting based classifiers with best performances. VotingClassifier1 (VC1) and VotingClassifier2(VC2) used only PM, D, SRL, POS and PI features. VotingClassifier3 (VC3) was also sampled with models that were trained on SM dimensionally reduced sets. The summary of the best models in our experiments presented in Table 6.

Table 6. Classification results of the best models.

Feature set	Classifier	AUC	Precision	Recall	F1-score
SM (PCA 10)	*RF*	*.74*	*.59*	*.71*	*.65*
–	*VC3*	*.74*	*.53*	*.81*	*.64*
UG	*RF*	*.72*	*.53*	*.74*	*.62*
–	*VC2*	*.72*	*.52*	*.77*	*.62*
SM (PCA 100)	*RF*	*.72*	*.71*	*.55*	*.62*
SM (PCA 5)	*AB*	*.72*	*.57*	*.68*	*.62*
PM	*XGB*	*.70*	*.50*	*.77*	*.61*
–	*VC1*	*.69*	*.49*	*.74*	*.59*

The evaluation revealed that SM feature set that based on Vkontakte users' subscriptions performed best on the data. The unigrams and psycholinguistic markers also demonstrated good results. We initially assumed that some of the psycholinguistic markers could work poorly on the data since users in our data usually write very short texts and the volume of concatenated posts cannot be compared to a logically connected text of the same size. This constrains are important for the specific of some psycholinguistic markers.

It should be noted that we can compare our results with the results of the Clef/eRisk 2017 Shared Task evaluation only with some restrictions. First, the language of Clef/eRisk 2017 was English, while our data is in Russian. Secondly, the number of data samples and the class ratio is different. Finally, depression class in Clef/eRisk 2017 Shared Task was assembled by manual expert examination of profiles from subforum related to the depression disorder. In our study, we operate only with the Beck Depression Inventory scores.

Despite this fact, the best F1-score reported in Clef/eRisk 2017 overview [29] was 64% achieved by the model that utilized tf-idf based features on the data with LIWC and dictionary features. In our experiments, tf-idf based features demonstrated 62% of F1-score with Random Forest classifier. It is important to mention, that the current state-of-art result on Clef/eRisk 2017 data is 73% of F1-score [30]. The best depression detection performance on our Vkontakte data is 65% of F1-score achieved by Random Forest classifier with dimensionally reduced subscription matrix.

6   Conclusion

In the study, we performed the depression detection task among 1330 users in Russian-speaking social network Vkontakte based on their text messages and profile information. By analyzing Beck Depression Inventory scores and processing the initial data we formed the set of 397 users' data samples with binary depression/control group labeling. We formed several text based feature sets including psycholinguistics markers, dictionaries, postags, semantic roles and relations, syntax relations, and n-grams. The general information from Vkontakte user profiles was used to form profile information features and subscription matrix.

The performed experiments revealed that information about social media subscriptions can be used to reveal depression with the best quality. It was also demonstrated that psycholinguistic markers, unigrams, and dictionaries features performed well on the data and can be effectively utilized for the depression detection task. The experiments were compared with Clef/eRisk 2017 Shared Task evaluation. The best result in our experiments is .65 of F1-score (.74 of ROC AUC score) achieved by the model that was trained on the subscription matrix transformed by PCA.

Thus, the analysis of depression linguistic markers and social media profile data is a promising area that can possibly make the prevention and treatment of depression more accessible to a large number of users. In the future work, we planning to examine neural network models for the depression detection task and evaluate regression analysis on the data using Beck Depression Inventory scores.

Acknowledgments. The reported study was funded by RFBR according to the research project 17-29-02225.

References

1. Turecki, G., Brent, D.A.: Suicide and suicidal behaviour. The Lancet **387**(10024), 1227–1239 (2016)
2. World Health Organization. https://www.who.int/mental_health/prevention/suicide/suicidepr event/en/. Accessed 19 Aug 2019
3. Belialov, F.I.: Depression, anxiety, stress, and mortality. Ter. Arkh. **88**(12), 116–119 (2016)
4. Surtees, P.G., Wainwright, N.W., Luben, R.N., Wareham, N.J., Bingham, S.A., Khaw, K.T.: Depression and ischemic heart disease mortality: evidence from the EPIC-Norfolk United Kingdom prospective cohort study. Am. J. Psychiatry **165**(4), 515–523 (2008)
5. Whang, W., et al.: Depression and risk of sudden cardiac death and coronary heart disease in women: results from the Nurses' Health Study. J. Am. Coll. Cardiol. **53**(11), 950–958 (2009)
6. Stankevich, M., Latyshev, A., Kuminskaya, E., Smirnov, I., Grigoriev, O.: Depression detection from social media texts. In: Elizarov, A., Novikov, B., Stupnikov, S. (eds.) Selected Papers of the XXI International Conference on Data Analytics and Management in Data Intensive Domains (DAMDID/RCDL 2019), vol. 2523, pp. 279–289. CEUR Workshop Proceedings, October 2019
7. Tausczik, Y.R., Pennebaker, J.W.: The psychological meaning of words: LIWC and computerized text analysis methods. J. Lang. Soc. Psychol. **29**(1), 24–54 (2010)
8. Kailer, A., Chung, C.K.: The Russian LIWC2007 dictionary. LIWC.net, Austin (2011)
9. Yates, A., Cohan, A., Goharian, N.: Depression and self-harm risk assessment in online forums. arXiv preprint arXiv:1709.01848 (2017)
10. Seabrook, E.M., Kern, M.L., Fulcher, B.D., Rickard, N.S.: Predicting depression from language-based emotion dynamics: longitudinal analysis of Facebook and Twitter status updates. J. Med. Internet Res. **20**(5), e168 (2018)
11. Al-Mosaiwi, M., Johnstone, T.: In an absolute state: elevated use of absolutist words is a marker specific to anxiety, depression, and suicidal ideation. Clin. Psychol. Sci. **6**(4), 529–542 (2018)
12. Panicheva, P., Ledovaya, Y., Bogolyubova, O.: Lexical, morphological and semantic correlates of the dark triad personality traits in Russian Facebook texts. In: 2016 IEEE Artificial Intelligence and Natural Language Conference (AINL), pp. 1–8. IEEE, November 2016
13. Bogolyubova, O., Panicheva, P., Tikhonov, R., Ivanov, V., Ledovaya, Y.: Dark personalities on Facebook: harmful online behaviors and language. Comput. Hum. Behav. **78**, 151–159 (2018)
14. Stankevich, M., Isakov, V., Devyatkin, D., Smirnov, I.: Feature engineering for depression detection in social media. In: ICPRAM, pp. 426–431 (2018)
15. Beck, A.T., Steer, R.A., Brown, G.K.: Beck depression inventory-II. San Antonio **78**(2), 490–498 (1996)
16. Losada, D.E., Crestani, F.: A test collection for research on depression and language use. In: Fuhr, N., et al. (eds.) CLEF 2016. LNCS, vol. 9822, pp. 28–39. Springer, Cham (2016). https://doi.org/10.1007/978-3-319-44564-9_3
17. MyStem. https://tech.yandex.ru/mystem. Accessed 19 Aug 2019
18. Straka, M., Straková, J.: Tokenizing, POS tagging, lemmatizing and parsing UD 2.0 with UDPipe. In: Proceedings of the CoNLL 2017 Shared Task: Multilingual Parsing from Raw Text to Universal Dependencies, pp. 88–99, August 2017
19. Koltsova, O.Y., Alexeeva, S., Kolcov, S.: An opinion word lexicon and a training dataset for Russian sentiment analysis of social media. In: Computational Linguistics and Intellectual Technologies: Materials of DIALOGUE 2016, Moscow, pp. 277–287 (2016)
20. Pennebaker, J.W.: The secret life of pronouns. New Sci. **211**(2828), 42–45 (2011)
21. Stankevich, M., Smirnov, I., Kuznetsova, Y., Kiselnikova, N., Enikolopov, S.: Predicting depression from essays in Russian. In: Computational Linguistics and Intellectual Technologies, DIALOGUE, vol. 18, pp. 637–647 (2019)

22. Devyatkin, D., Kuznetsova, Y., Chudova, N., Shvets, A.: Intellectual analysis of the manifestations of verbal aggressiveness in the texts of network communities [Intellektuanyj analiz proyavlenij verbalnoj agressivnosti v tekstah setevyh soobshchestv]. Artif. Intell. Decis. Making **2**, 27–41 (2014)

23. Droganova, K., Lyashevskaya, O., Zeman, D.: Data conversion and consistency of monolingual corpora: Russian UD treebanks. In: Proceedings of the 17th International Workshop on Treebanks and Linguistic Theories (TLT 2018), Oslo University, Norway, 13–14 December 2018, no. 155, pp. 52–65. Linköping University Electronic Press (2018)

24. Shelmanov, A.O., Smirnov, I.V.: Methods for semantic role labeling of Russian texts. In: Computational Linguistics and Intellectual Technologies: Papers from the Annual International Conference "Dialogue", vol. 13, no. 20, pp. 580–592 (2014)

25. Pedregosa, F., et al.: Scikit-learn: machine learning in Python. J. Mach. Learn. Res **12**(1), 2825–2830 (2011)

26. Chen, T., Guestrin, C.: XGBoost: a scalable tree boosting system. In: Proceedings of the 22nd ACM SIGKDD International Conference on Knowledge Discovery and Data Mining, pp. 785–794, August 2016

27. Ke, G., et al.: LightGBM: a highly efficient gradient boosting decision tree. In: Advances in Neural Information Processing Systems, pp. 3146–3154 (2017)

28. Wold, S., Esbensen, K., Geladi, P.: Principal component analysis. Chemometr. Intell. Lab. Syst. **2**(1–3), 37–52 (1987)

29. Losada, D.E., Crestani, F., Parapar, J.: eRISK 2017: CLEF lab on early risk prediction on the internet: experimental foundations. In: Jones, G.J.F., et al. (eds.) CLEF 2017. LNCS, vol. 10456, pp. 346–360. Springer, Cham (2017). https://doi.org/10.1007/978-3-319-65813-1_30

30. Trotzek, M., Koitka, S., Friedrich, C.M.: Utilizing neural networks and linguistic metadata for early detection of depression indications in text sequences. IEEE Trans. Knowl. Data Eng. (2018)

Tatar WordNet: The Sources
and the Component Parts

Alexander Kirillovich[1(✉)], Alfiya Galieva[2], Olga Nevzorova[1], Marat Shaekhov[2],
Natalia Loukachevitch[1], and Dmitry Ilvovsky[1]

[1] Kazan Federal University, Kazan, Russia
alik.kirillovich@gmail.com, onevzoro@gmail.com, louk_nat@mail.ru,
dilvovsky@hse.ru
[2] Tatarstan Academy of Sciences, Kazan, Russia
amgalieva@gmail.com, q-mir-bey@list.ru

Abstract. We describe an ongoing project of construction of the Tatar Wordnet. The Tatar Wordnet is being constructed on the base of three source resources, developed by us. The first source is TatThes, a bilingual Russian-Tatar Social-Political Thesaurus. TatThes, in turn, has been constructed by manual translation and extension of RuThes, a linguistic ontology for Russian. The second source is a Tatar translation of RuWordNet, a wordnet for Russian. This translation was carried out automatically on the base of a Russian-Tatar dictionary, and then was manually verified. The third source is a semantic classification of Tatar verbs, developed from scratch. We discuss the structure, methodology of compilation and the current state these source resources, and justify the choice of them as the initial resources for building the Tatar Wordnet. Our ultimate goal is to publish Tatar Wordnet on the Linguistic Linked Open Data cloud and integrate it to the Global WordNet Grid.

Keywords: Tatar language · WordNet · Linguistic Linked Open Data

1 Introduction

The Princeton WordNet thesaurus (PWN) [1, 2] is one of the most important language resources for linguistic studies and natural language processing. PWN is a large-scale lexical knowledge base for English, organized as a semantic network of synsets. A synset is a set of words with the same part-of-speech that can be interchanged in several contexts. Synsets are interlinked by semantic relations, such as hyponymy (between specific and more general concepts), meronymy (between parts and wholes), antonymy (between opposite concepts) and other.

Inspired by success of PWN, many projects have been initiated to develop wordnets for other languages across the globe. Nowadays wordnet-like resources are developed for nearly 80 languages, but Tatar language is not among them. To fill this gap, we lunched a project of construction TatWordNet, a wordnet-like resource for Tatar.

There are two main approaches for construction of wordnets for new languages: expand and merge [3]. The expand approach is to take the semantic network of PWN

© Springer Nature Switzerland AG 2020
A. Elizarov et al. (Eds.): DAMDID/RCDL 2019, CCIS 1223, pp. 195–208, 2020.
https://doi.org/10.1007/978-3-030-51913-1_13

and translate its synsets into the target language, adding additional synsets when needed. The merge approach is to develop a semantic network in the target language from scratch and then link it to PWN.

Since the merge approach is very labor-intensive and time consuming, the expand approach seems more appropriate for under-resources languages such as Tatar. However, in development of Tatar WordNet, the expand approach can't be directly applied either, due to the lack of large English-Tatar dictionaries, necessary for translation of PWN to Tatar. At the same time, there are several relatively large and high-quality Russian-Tatar dictionaries, so Russian thesauri can be used as the source resources instead of PWN.

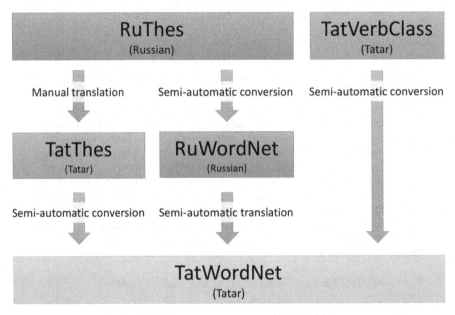

Fig. 1. The source resources of TatWordNet

With this consideration in mind we are constructing TatWordNet on the base of three source resources, developed by us (Fig. 1). The first source is TatThes, a bilingual Russian-Tatar Social-Political Thesaurus. TatThes, in turn, has been constructed by manual translation and extension of RuThes, a linguistic ontology for Russian. The second source is a Tatar translation of RuWordNet, a wordnet for Russian. RuWordNet, for its part, has been constructed by semi-automatic conversion of RuThes. The translation of RuWordNet to Tatar was carried out automatically on the base of a Russian-Tatar dictionary, and then was manually verified. The third source is a semantic classification of Tatar verbs, developed from scratch.

In this paper, we describe the methodology for constructing TatWordNet, and the source resources used in this constructing. The paper is an extended version of our short paper [4], and describes processing of all the source resources (TatThes, Tatar translation of RuWordNet and TatVerbClass), while in [4] processing of the only one source has been described.

The rest of the paper is organized as follows. Section 2 outlines the basic theoretical background of the study, and the main attention is paid to wordnet projects developed for the Turkic languages. Section 3 presents the methodology of compiling the Russian-Tatar socio-political thesaurus and its current state. Section 4 describes the most important aspects of implementing a wordnet-like resource using Tatar thesaurus synsets for Tatar nouns. Section 5 describes a Tatar translation of RuWordNet, and Sect. 6 describes a semantic classification of Tatar verbs. Section 7 discusses the conclusions and outlines the prospects of future work.

2 Related Works

At present time, there are various wordnets for some Turkic languages.

Two Turkish wordnet projects have been developed for the Turkish language. The first one [5, 6] has been created at Sabancı University as part of the BalkaNet project [7]. The BalkaNet project was built on the basis of a combination of expand and merge approaches. All wordnets contain many synonyms for Balkan common topics, as well as synsets typical for each of the BalkaNet languages. The size of Turkish Wordnet is about 15,000 synsets.

Another Turkish wordnet is the KeNet [8, 9]. This wordnet was built on the basis of modern Turkish dictionaries. To build this resource, a bottom-up approach was used. Based on dictionaries, words were selected and then manually grouped into synsets. The relationships between words have been automatically extracted from dictionary definitions and then the latter have been fixed between synsets. The size of this resource is about 113,000 synsets.

Unfortunately, lack of large Turkish-Tatar dictionaries (as well as English-Tatar ones) makes it impossible to translate Turkish resources into the Tatar language. In this respect, the Tatar language can be attributed to low-resource languages.

The Extended Open Multilingual Wordnet [10] resource is built from Open Multilingual Wordnet by replenishing the WordNet data automatically extracted from the Wiktionary and Unicode Common Locale Data Repository (CLDR). The resource contains wordnets for 150 languages, including several Turkic: Azerbaijani, Kazakh, Kirghiz, Tatar, Turkmen, Turkish, and Uzbek. The Tatar wordnet contains a total of 550 concepts, which covers 5% of the PWN core concepts.

The BabelNet [11] resource contains a common network of concepts that have text inputs in many languages. The BabelNet contains 90,821 Tatar text entries that refer to 63,989 concepts. However, due to the fact that this resource was built automatically, it has quality issues.

Thus, the development of a quality Tatar wordnet with an emphasis on the specific features of the Tatar language based on the existing lexical resources is very relevant.

3 Tatar Socio-Political Thesaurus: Methodological Issues of Compiling and Current State

The conceptual model of the Tatar socio-political thesaurus (hereinafter referred to as TatThes) and the general principles of displaying linguistic data are taken from the RuThes project (http://www.labinform.ru/pub/ruthes/) [12, 13]. The RuThes thesaurus is a hierarchical network of concepts with attributed lexical entries for automatic text processing.

In RuThes each concept is linked with a set of language expressions (nouns, adjectives, verbs or multiword expressions of different structures – noun phrases and verb phrases) which refer to the concept in texts (lexical entries). RuThes concepts have no internal structure as attributes (frame elements), so concept properties are described only by means of relations with other concepts.

Each of RuThes concepts is represented as a set of synonyms or near-synonyms (plesionyms). RuThes developers use a weaker term, ontological synonyms, to designate words belonging to different parts of speech (like stabilization, to stabilize); the items may be related to different styles and genres. Ontological synonyms are the most appropriate means to represent cross-linguistic equivalents (correspondences), because such approach allows us to fix units of the same meaning disregarding surface grammatical differences between them. For example, Table 1 represents basic ways of translating Russian adjective + noun phrases into Tatar.

Table 1. Examples of Russian A*dj* + *Noun* phrases and ways of translating them into Tatar

Russian unit	Corresponding Tatar unit	The structure of Tatar unit	English translation
Пенсионный возраст	Пенсия яше	$N + N_{POSS\_3}$	Retirement age
Рабочий класс	Эшчелэр сыйныфы	$N_{PL} + N_{POSS\_3}$	Working class
Консульская служба	Консуллык хезмэте	$N_{NMLZ} + N_{POSS\_3}$	Consular service
Сексуальное меньшинство	Сексуаль азчылык	$ADJ + N$	Sexual minority
Именная стипендия	Исемле стипендия	$N_{COMIT} + N_{PL}$	Nominal scholarship

TatThes is based on the list of concepts of RuThes, i.e. the Tatar component is based on the list of concepts of the RuThes thesaurus. The methodology of compiling the Tatar part of the thesaurus includes the following steps:

1. Search for equivalents (corresponding words and multiword expressions) which are actually used in Tatar as translations of Russian items.
2. Adding new concepts representing topics which are important for the sociopolitical and cultural life of the Tatar society and which are not presented in the original RuThes (for example, Islam-related concepts, designations of Tatar culture specific phenomena, etc.).

3. Revising relations between the concepts considering the place of each new concept in the hierarchy of the existing ones and, if necessary, adding new concepts of the intermediate level. So an important step is to check up the parallelism of conceptual structures between the languages.

TatThes is mainly being compiled by manual translation of terms from RuThes into Tatar; besides the Tatar language specific concepts and their lexical entries are added (about 250 new concepts). Search for equivalents in the Tatar language in many cases became a time-consuming task because available Russian-Tatar dictionaries of general purpose contain obsolete lexical data [14]. So when compiling the lists of concept names and lexical entries we manually browsed large arrays of official documents and media texts in Tatar. In the process of compiling the Thesaurus, data from the following available Tatar corpora is used:

1. Tatar National Corpus (http://tugantel.tatar/?lang=en);
2. Corpus of Written Tatar (http://www.corpus.tatar/en).

In the course of the project, we found that a distinguishing feature of contemporary Tatar lexicon is a great deal of absolute synonyms of different origin and structure, the main cause of the phenomenon being language contacts [15].

TatThes is implemented as a web application and has a special site (http://tattez. turklang.tatar/). Additionally, it has been published in the Linguistic Linked Open Data cloud as part of RuThes Cloud project [16]. Currently TatThes contains 10,000 concepts, 6,000 of them provided with lexical entries.

4 Tatar Thesaurus Data for Wordnet Implementation: Case of Nouns

Previously, the RuThes thesaurus has been semi-automatically converted to a wordnet-like structure, and a Russian wordnet (RuWordNet) has been generated [17, 18]. The conversion included two main steps:

1. automatic subdivision of RuThes text entries into three nets of synsets according to parts of speech;
2. semi-automatic conversion of RuThes relations to wordnet-like relations.

The current version of RuWordNet (http://ruwordnet.ru/eng) contains 110 thousand Russian unique words and expressions. The same approach can be used to transform TatThes to Tatar wordnet.

The TatThes data may serve as an initial basis for wordnet building for the following reasons:

1. The sociopolitical sphere covers a broad area of modern social relations. This area comprises generally known terms of politics, international relations, economics and finance, technology, industrial production, warfare, art, religion, sports, etc.

2. Currently TatThes, in addition to terminology, comprises some general lexicon branches representing lexical items which can be found in various domain specific texts.
3. Semantic relations in TatThes are necessary and sufficient to arrange the Tatar nominal vocabulary (nouns and noun phrases) as a wordnet-like network of synsets.

Thesaurus concepts unite synonymous items, so we have ready sets of synonyms as building blocks for the wordnet. The concepts are linked by semantic relations with each other. In the RuThes and in the TatThes there are four main types of relationships between concepts, see Table 2. Semantic relations, mapped in the wordnet, are not shared by all lexical categories, so converting thesaurus data into the wordnet format requires dissimilar ways for different parts of speech.

Table 2. Semantic relations between nouns in the thesaurus and in wordnets

Semantic relations in the Thesaurus	Semantic relations in wordnets
Hypernym—hyponyms	Hypernym—hyponyms
Holonym—meronym	Holonym—meronym
Symmetrical association (Asc)	
Asymmetric association (Asc1/Asc2)	

Asc and Asc1/Asc2 association relations need additional explanations. The Asc symmetrical association, distinguished in RuThes and inherited by the Tatar Socio-Political Thesaurus, connects very similar concepts, which the developers did not dare to combine into the same concept (for example, cases of presynonymy of items).

The Asc1/Asc2 asymmetric association connects two concepts that cannot be described by the relations mentioned above, but neither of them could exist without the other (for example, concept SUMMIT MEETING needs existence of the concept HEAD OF THE STATE). In studies of ontologies this relation may be mapped as the ontological dependence relation.

Nevertheless, basic semantic relations which we need to group noun concepts into the wordnet are presented in TatThes.

The core of TatThes is made up of nouns and noun phrases (see Table 3), so the bulk of thesaurus data may be used for Tatar wordnet building without significant changes (synonymous items are yet joined into synsets and the required relations between them are selected).

An important issue is reflecting Tatar specific word usage features in the resource. Mere presence of the shared concepts in languages does not necessarily evidence the same ways of usage of individual words or of usage of words of individual semantic classes. Consider this in the following example. A specific feature of the Tatar language is using hypernyms before a corresponding hyponym, and such usage is not regarded as pleonasm in many cases (examples 1–3):

Table 3. Number of noun concepts and noun phrase concepts in TatThes (on the data of the Russian part)

Structure of TatThes items	Number of items
Noun	3387
Adj + Noun	3135
Noun + Noun$_{GEN}$	352
Other	3126
Total	10000

(1) *Париж шәһәрендә* 'in the city of Paris' (instead of 'in Paris');
(2) *кыз кеше* 'girl human' (instead of 'a girl');
(3) *май аенда* 'in the month of May' (instead of 'in May').

In cases when such usage is conventionalized and corpus data evidences that the usage has a high frequency, we include such hyponym-hypernym items into the list of lexical entries of the concept. Such manner of designating is a feature of using toponyms and some classes of general lexicon, so it should be considered in Tatar wordnet building. For example, lexical entries of month names include such conventionalized noun phrases, composed of the month name and the hyponym designating month in general, see Table 4.

Table 4. Representing lexical entries of month names in the Thesaurus

Russian concept name	Russian lexical entries	Rus POS	Tatar concept name	Tatar lexical entries	Tat POS
ДЕКАБРЬ	Декабрь 'December' Декабрьский 'of December'	N ADJ	Декабрь	Декабрь 'December' Декабрь ае 'month of December'	N NP
ЯНВАРЬ	Январь 'January' Январский 'of January'	N ADJ	Гыйнвар	Гыйнвар 'January' Гыйнвар ае 'month of January' Январь 'January' Январь ае 'month of January'	N NP N NP
ФЕВРАЛЬ	Февраль 'February' Февральский 'of February'	N ADJ	Февраль	Февраль 'February' Февраль ае 'month of February'	N NP

Because RuThes concepts assemble ontological synonyms, RuThes lexical entries bring together words of different parts of speech. Therefore in standard cases a Russian synset joins a noun (often we use it as a concept name) and a relative adjective derived from the noun (Table 5; only core items of synsets are represented). In Tatar, like in other Turkic languages, there are no original relative adjectives (and existing ones are borrowed from European or Oriental languages), so in many cases TatThes synsets are composed of items of the same part of speech, mainly of nouns. This circumstance greatly facilitates cleaning thesaurus synsets data for wordnet developing.

Table 5. Typical arrangement of Russian and Tatar Thesaurus synsets

Basic lexical entries of Russian concept	Part of speech of Russian words	Basic lexical entries of Tatar concept	Part of speech of Tatar words
Река 'river'	N	Елга 'river'	N
Речной 'of river, fluvial'	ADJ		
Факультет 'faculty'	N	Факультет 'faculty'	N
Факультетский 'of faculty'	ADJ		
Преподаватель 'teacher'	N	Укытучы 'teacher'	N
Преподавательский 'of teacher'	ADJ		
Больница 'hospital'	N	Хастаханә 'hospital'	N
Больничный 'of hospital'	ADJ	Сырхауханә 'hospital'	N

So the core of TatThes is made up of nouns and noun phrases (69% of total number of concepts). At the moment semantic relations between nouns mapped in the thesaurus, are necessary and sufficient to convert the Tatar thesaurus data into the wordnet format.

5   Tatar Translation of RuWordNet

In this section we describe Tatar translation of RuWordNet.

At first stage we performed automatic translation of RuWordNet resource with the help of the Russian-Tatar dictionary edited by F.A. Ganiev.

The next main task was manual verification of automatically obtained data. Using the data on hyponyms and hyperonyms, as well as the glossary, we checked the word meaning since the priority was not to evaluate the correct translation of individual words, but to the translation of the concepts of the original words into the target language. By analyzing and editing the text input in the Tatar language, one can see the following language situations (cases):

1) Noun synsets in the Russian language contain items derived from words of different parts of speech, for example, deverbal nouns naming actions and states. Words of different meaning and derivation models may be translated into Tatar differently. For example, often Russian deverbal nouns are conveyed in Tatar as verbal nouns – a hybrid grammatical class sharing some features of nouns and verbs (verbal nouns are the standard way to fix verbs in Tatar dictionaries). As a result, Russian noun synsets may correspond to Tatar synsets contacting items of dissimilar grammatical classes:

— *величие* (ru) 'greatness' - *бөеклек, олылык* (tat) – nouns in both languages;
— *бездействие* (ru) 'inaction' - *бер нәрсә дә эшләмәу* (one + what + PART + do-NEG, VN), *чара күрмәу* (measure + see-NEG, VN) (tat) – a noun in Russian and phrases with a verbal noun as a node word in Tatar;
— *гегемония* (ru) 'hegemony'/ - *гегемония, җитәкчелек иту* (leader-NMLZ + do), *өстенлек* (tat) – in the Tatar parts are nouns and a verbal noun phrase;
— *вескость* (ru) 'weightiness, validity' - *авыр булу* (heavy +be-VN), *саллы булу* (weighty + be-VN) (tat) - verbal noun phrases in Tatar.

Here and in examples below only the Tatar items with the grammatical structure differing from Russian correspondences are glossed.

2) There are many words (about 375) translated into the Tatar language by using a descriptive construction because these words are not presented in Tatar dictionaries:

— *коренник* (ru) 'shaft-horse'- *төпкә җигелгән ат* (bottom-DIR + harness- PASS, PCP_PS + horse) (tat);
— *выскочка (ru)* 'upstart' - *сикергәк, ялагай, сәнәктән көрәк булган кеше* (pitchfork-ABL + shovel + be-PCP_PS + man) (tat) – nouns and a noun phrase in Tatar.

Such descriptive phrases can be divided into 4 groups, depending on the lexical meaning and source word parts:

A) Root words that do not have a corresponding version in the Tatar language due to the fact that these concepts are not characteristic of the mode of life and the culture of this people. E.g.,

— *именинник* (ru) 'birthday boy' - *исем бәйрәмен* (name + fete-POSS-3, ACC + perform-VN, PCP_PR + man) (tat);
— *клюка (*ru) 'crooked top stick' – *кәкре башлы таяк* (crooked + head-ATTR_MUN + stick) (tat);

B) Terms and concepts that do not have equivalents in the Tatar language, transferred borrowed-words and/or descriptive phrases: *дротик* (ru) 'dart' – *дротик*, 'dart' – *кыска саплы сөнге* (short + handle-ATTR_MUN + spear) (tat).

C) Compound words that do not have equivalents identical in structure in the target language. E.g. many Russian two root words are conveyed in Tatar by means of compounds:

— *водосток* (ru) 'gutter' – *су агып төшә торган торба* (water + pour-CONV +go down-PCP_PR + be-PCP_PS + tube) (tat);
— *двустволка* (ru) 'shotgun' – *ике көпшәле мылтык* (two +barrel- ATTR_MUN + gun) (tat);
— *естествоиспытатель* 'naturalist' – *табигать фәннәре белгече* (nature + science-PL, POSS_3, specialist-POSS_3) (tat).

D) Many Tatar synsets contain in addition phrases with a hypernym, in particular, names of months, plants, trees, nationalities, and other classes:

— *январь* (ru) 'January' – *гыйнвар, гыйнвар ае* (January + month-POSS_3) (tat);
— *вяз* (ru) 'elm'– *карама, карама агачы* (elm + tree-POSS_3) (tat);
— *липа* (ru) 'linden' – *юкэ, юкэ агачы* (linden+ tree-POSS_3) (tat);
— *японец* (ru) 'Japanese man' – *япон, япон кешесе* (Japanese + man-POSS_3) (tat);
— *девочка (ru)* 'female child' – *кыз бала* (girl + child) (tat);
— *иноходец* (ru) 'pacer horse' – *юрга, юрга ат* (pacer + horse) (tat).

3) The Tatar language has no morphological category of grammatical gender, and to convey this category, lexical means are used. So in Tatar synsets corresponding to Russian synsets gathering words denoting females, words specifying the age and the marital status is added to such text entries for the Tatar language (*кыз* 'girl' or *хатын* 'woman'). This applies to translation names of nationalities, professions, social status, etc.:

— *активистка* (ru) 'activist woman or girl' – *активистка, актив хатын, актив кыз*;
— *караимка* (ru) 'karaite' woman or girl – *караим хатыны, караим кызы*;
— *купальщица* (ru) bather woman or girl' – *су коенучы хатын, су коенучы кыз*;
— *манекенщица* (ru) 'mannequin, fashion model woman or girl' – *манекенчы хатын, манекенчы кыз*.

4) As we mentioned above, a problematic area to translate is synsets in Russian for which there are no corresponding concepts in the Tatar culture. A significant portion of them make up the concepts of Orthodox Christianity absent in Islam (the latter is the religion of the most part of Tatars). We found currently 32 such items. For example:

— *ересь* (ru) 'heresy' – *ересь* (tat);
— *молебен* (ru) 'prayer service' – *молебен* (tat);
— *миропомазание* (ru) 'anointing'– *миро белэн майлап чукындыру* (chrism + with + oil-CONV + baptize-VN) (tat).

Religious items are translated in three ways:

A) by using words borrowed from Russian (however the origin of words may be different, for example Greek);
B) by using explanatory translation;
C) by using words denoting close concepts from the Muslim terminology.

6　Database of Semantic Classes of Verbs

In this section, we describe TatVerbClass, a database of semantic classes of Tatar verbs [19].

The classification scheme is based on the following parameters of verbal lexemes:

1. thematic feature, linked with the verb's thematic class, which allows us to mark up the verb's denotation sphere;

2. grammatical feature, linked with the valency changing operations of voice affixes (possibility of producing grammatical voice derivatives and particular meanings of voice forms);
3. syntactic feature, related to the allowable predicate-argument structure and thematic roles of arguments;
4. derivational feature, related to the verb's derivation pattern (grammatical class of the stem, derivational meaning of the verb forming affix).

Each verb is provided with a semantic tag (or with a set of the latter), there have been distinguished 59 basic semantic (ontological) classes, such as movement verbs, speech verbs, etc.). Semantic classes may join items with dissimilar individual meanings, grammatical properties, syntactic behavior, etc. So in a semantic class we distinguish a set of individual subclasses including verbs of similar structure, features and behavior.

In spite of rather formal criteria when selecting subclasses (ability to produce the same grammatical voice derivatives and sharing argument structure of verbs are in the foreground), the words of similar meaning fall into the same subclass. In most cases subclasses join synonyms, antonyms and hyponyms related to the same hypernym (see Tables 6, 7 – examples of subclasses of the physiological verbs).

Table 6. Subclass of verbs related to the hypernym 'to feel sensations in the body'

Verbs	English translation (the main senses)	Thematic tags in DB
авырт	to ache (on physical pain)	t:physiol, t:perc
сызла	to ache (on intensive pain)	t:physiol, t:perc
әрне	to ache (on acute pain)	t:physiol, t:perc
ачыт	to ache (on burning pain)	t:physiol, t:perc
кычыт	to itch and tickle	t:physiol, t:perc
кызыш	to feel fever	t:physiol, t:perc
кымыржы	to itch	t:physiol, t:perc
чымырда	to feel goosebumps	t:physiol, t:perc
чемердә	to feel goosebumps	t:physiol, t:perc

All the verbs in Table 6 share the features:

– all the verbs have a basic meaning 'to feel some sensations in the body/part of the body' and they are provided with the same semantic tags;
– all the verbs are intransitive and express a state;
– all the verbs may have causative derivatives and can not produce passive and reciprocal derivatives;
– as a standard syntactic subject they have nouns denoting body or parts of body.

Table 7. Verbs denoting disease states

Verbs	English translation (the main senses)	Thematic tags in DB
авыр	to be ill	t:physiol:disease
чирлә	to be ill	t:physiol:disease
сырхала	to be ill	t:physiol:disease
хастала	to be ill	t:physiol:disease

Another example is a subclass of verbs denoting disease states (Table 7), where all the items are synonyms.

The verbs самсыра 'to ill, be down at health' despite the semantic affinity with the verbs from Table 7, is set outside the scope of the subclass, because it does not take arguments with белән 'with' postposition, unlike the verbs represented in the Table 7.

7   Conclusion

In this paper, we described the methodology for constructing TatWordNet on the base of the three resources: Russian-Tatar Social-Political Thesauru (TatThes), Tatar translation of RuWordNet and the Database of Semantic Classes of Tatar Verbs (TatVerbClass). Currently, TatWordNet consists in the three components, corresponding to these sources. Our immediate goal is to complete development of these components and merge them into single unified resource.

After that we are planning to continue our research in the following directions:

1. checking the quality and representativeness of the data obtained through comparison with frequency dictionary created on the basis of the "Tugan tel" Tatar National corpus and adding missing senses;
2. comparing the core data of TatWordNet with the core of Princeton WordNet and adding missing senses;
3. developing hierarchies for adjectives and other parts-of-speech.

Our ultimate goal is to publish Tatar Wordnet on the Linguistic Linked Open Data cloud [20] and integrate it to the Global WordNet Grid [21] via the Collaborative Interlingual Index.

Acknowledgments. This work was funded by Russian Science Foundation according to the research project no. 19-71-10056.

References

1. Fellbaum, C. (ed.): WordNet: An Electronic Lexical Database. MIT Press, Cambridge (1998)

2. Fellbaum, C.: WordNet. In: Poli, R., et al. (eds.) Theory and Applications of Ontology: Computer Applications, pp. 231–243. Springer, Heidelberg (2010). https://doi.org/10.1007/978-90-481-8847-5_10

3. Vossen, P. (ed.): EuroWordNet: General Document. University of Amsterdam (2002). http://www.illc.uva.nl/EuroWordNet/docs.html

4. Galieva, A., Kirillovich, A., Loukachevich, N., and Nevzorova, O.: Towards a Tatar WordNet: a methodology of using tatar thesaurus. In: Elizarov, A., et al. (eds.) Selected Papers of the XXI International Conference on Data Analytics and Management in Data Intensive Domains (DAMDID/RCDL 2019). CEUR Workshop Proceedings, vol. 2523, pp. 316–324. CEUR-WS (2019)

5. Çetinoğlu, Ö., Bilgin, O., Oflazer, K.: Turkish Wordnet. In: Oflazer, K., Saraçlar, M. (eds.) Turkish Natural Language Processing. TANLP, pp. 317–336. Springer, Cham (2018). https://doi.org/10.1007/978-3-319-90165-7_15

6. Bilgin, O., Çetinoğlu, Ö., Oflazer, K.: Building a Wordnet for Turkish. Romanian J. Inf. Sci. Technol. 7(1–2), 163–172 (2004)

7. Tufis, D., Cristea, D., Stamou, S.: BalkaNet: aims, methods, results and perspectives. A general overview. Romanian J. Inf. Sci. Technol. 7(1–2), 9–43 (2004)

8. Ehsani, R.: KeNet: a comprehensive Turkish wordnet and using it in text clustering. Ph.D. thesis. Işık University (2018)

9. Ehsani, R., Solak, E., Yildiz, O.T.: Constructing a WordNet for Turkish using manual and automatic annotation. ACM Trans. Asian Low-Resource Lang. Inf. Process. 17(3), Article No. 24 (2018). https://doi.org/10.1145/3185664

10. Bond, F., Foster, R.: Linking and extending an Open Multilingual WordNet. In: Schuetze, H., Fung, P., Poesio, M. (eds.) Proceedings of the 51st Annual Meeting of the Association for Computational Linguistics (ACL 2013), Volume 1: Long Papers, pp. 1352–1362. ACL (2013)

11. Navigli, R., Ponzetto, S.P.: BabelNet: the automatic construction, evaluation and application of a wide-coverage multilingual semantic network. Artif. Intell. 193, 217–250 (2012). https://doi.org/10.1016/j.artint.2012.07.001

12. Loukachevich, N., Dobrov, B.: RuThes linguistic ontology vs. Russian wordnets. In: Orav, H., Fellbaum, C., Vossen, P. (eds.) Proceedings of the 7th Conference on Global WordNet (GWC 2014), pp. 154–162. University of Tartu Press (2014)

13. Loukachevich, N.V., Dobrov, B.V., Chetviorkin, I.I.: RuThes-Lite, a publicly available version of thesauru of Russian language RuThes. In: Computational Linguistics and Intellectual Technologies: Papers from the Annual International Conference "Dialogue", pp. 340–349. RGGU (2014)

14. Galieva, A., Kirillovich, A., Khakimov, B., Loukachevich, N., Nevzorova, O., Suleymanov, D.: Toward domain-specific Russian-Tatar thesaurus construction. In: Bolgov, R., et al. (eds.) Proceedings of the International Conference on Internet and Modern Society (IMS-2017), pp. 120–124. ACM Press (2017). https://doi.org/10.1145/3143699.3143716

15. Galieva, A., Nevzorova, O., Yakubova, D.: Russian-Tatar socio-political thesaurus: methodology, challenges, the status of the project. In: Mitkov, R., Angelova, G. (eds.) Proceedings of the International Conference Recent Advances in Natural Language Processing (RANLP 2017), pp. 245–252. INCOMA Ltd. (2017). https://doi.org/10.26615/978-954-452-049-6_034

16. Kirillovich, A., Nevzorova, O., Gimadiev, E., Loukachevich, N.: RuThes Cloud: towards a multilevel linguistic linked open data resource for Russian. In: Różewski, P., Lange, C. (eds.) KESW 2017. CCIS, vol. 786, pp. 38–52. Springer, Cham (2017). https://doi.org/10.1007/978-3-319-69548-8_4

17. Loukachevich, N.V., Lashevich, G., Gerasimova, A.A., Ivanov, V.V., Dobrov, B.V.: Creating Russian WordNet by conversion. In: Computational Linguistics and Intellectual Technologies: Papers from the Annual Conference "Dialogue", pp. 405–415. RGGU (2016)

18. Loukachevitch, N., Lashevich, G., Dobrov, B.: Comparing two thesaurus representations for Russian. In: Bond, F., Kuribayashi, T., Fellbaum, C., Vossen, P. (eds.) Proceedings of the 9th Global WordNet Conference (GWC 2018), pp. 35–44. GWA (2018)
19. Galieva, A., Vavilova, Z., Gatiatullin, A.: Semantic classification of Tatar verbs: selecting relevant parameters. In: Čibej, J., Kosem, I., and Krek, S. (eds.) Proceedings of the XVIII EURALEX International Congress: Lexicography in Global Contexts (Euralex 2018), pp. 811–818. Ljubljana University Press (2018)
20. Cimiano, P., Chiarcos, C., McCrae, J.P., Gracia, J.: Linguistic Linked Open Data Cloud. In: Cimiano, P., et al. (eds.) Linguistic Linked Data: Representation, Generation and Applications, pp. 29–41. Springer, Cham (2020). https://doi.org/10.1007/978-3-030-30225-2_3
21. Vossen, P., Bond, F., McCrae, J.P.: Toward a truly multilingual Global Wordnet Grid. In: Barbu Mititelu, V., et al. (eds.) Proceedings of the 8th Global WordNet Conference (GWC 2016), pp. 419–426. GWA (2016)

Distributed Computing

An Approach to Fuzzy Clustering of Big Data Inside a Parallel Relational DBMS

Mikhail Zymbler$^{(\boxtimes)}$ ⓘ, Yana Kraeva, Alexander Grents, Anastasiya Perkova, and Sachin Kumar ⓘ

South Ural State University, Chelyabinsk, Russia
{mzym,kraevaya,grentsav,perkovaai}@susu.ru,
sachinagnihotri16@gmail.com

Abstract. Currently, despite the widespread use of numerous NoSQL systems, relational DBMSs remain the basic tool for data processing in various subject domains. Integration of data mining methods with relational DBMS is a topical issue since such an approach avoids export-import bottleneck and provides the end-user with all the built-in DBMS services. Proprietary parallel DBMSs could be a subject for integration of data mining methods but they are expensive and oriented to custom hardware that is difficult to expand. At the same time, open-source DBMSs are now being a reliable alternative to commercial DBMSs and could be seen as a subject to encapsulate parallelism. In this study, we present an approach to fuzzy clustering of very large data sets inside a PDBMS. Such a PDBMS is obtained by small-scale modifications of the original source code of an open-source serial DBMS to encapsulate partitioned parallelism. The experimental evaluation shows that the proposed approach overtakes parallel out-of-DBMS solutions with respect to export-import overhead.

Keywords: Big Data · Parallel DBMS · PostgreSQL · Fuzzy clustering

1 Introduction

Currently, despite the widespread use of numerous NoSQL systems, relational DBMSs remain the basic tool for data processing in various subject domains. To make sure of this fact, let us take a look to statistics from the DB-Engines.com portal. At the end of 2019, relational DBMSs kept the last place by popularity among the data management systems (by their mentions in news feeds, social and professional networks, etc.) yielding all the NoSQL systems. At the same time, ranking of the systems above by database model showed that relational DBMSs take three quarters of the market.

In addition to OLTP and OLAP scenarios, current data intensive applications should provide data mining abilities. This fact makes the integration of data mining methods with relational DBMS a topical issue [18]. Indeed, if we

ⓒ Springer Nature Switzerland AG 2020
A. Elizarov et al. (Eds.): DAMDID/RCDL 2019, CCIS 1223, pp. 211–223, 2020.
https://doi.org/10.1007/978-3-030-51913-1_14

consider DBMS only as a fast and reliable data repository, we get significant overhead for export large data volumes outside a DBMS, changing data format, and import results of analysis back into a DBMS. Moreover, such an integration provides the end-user with all the built-in DBMS services (query optimization, data consistency and security, etc.).

Nowadays, Big Data phenomenon demands parallel DBMSs (PDBMSs) to processing very large databases. Proprietary PDBMSs could be a subject for integration of data mining methods but they are expensive and oriented to custom hardware that is difficult to expand. Open-source DBMSs are now being a reliable alternative to commercial DBMSs but there is a lack of open-source parallel DBMSs since the development of such a complex software system is rather expensive and takes a lot of time.

This paper is a revised and extended version of the invited talk [31]. In the paper, we present an approach to fuzzy clustering of very large data sets inside a PDBMS. Such a PDBMS is obtained by small-scale modifications of the original source code of an open-source serial DBMS to encapsulate partitioned parallelism. The paper is structured as follows. Section 2 contains a short overview of related work. Section 3 describes a method of encapsulation of partitioned parallelism into serial DBMS. In Sect. 4, we present fuzzy clustering algorithm for PDBMS described above. In Sect. 5, we give the results of the experimental evaluation of our approach. Finally, Sect. 6 summarizes the results obtained and suggests directions for further research.

2   Related Work

Research on the integration of data analytic methods with relational DBMS started as far as data mining became full-fledged scientific discipline. First investigators proposed mining query languages [7,10] and SQL extensions for data mining [15,29]. There are SQL implementations of algorithms to solve data mining basic problems, namely association rules [26,27], classification [19,25], and clustering [16,17], as well as graph mining problems [14,21].

The MADlib library [8] provides many data mining methods inside PostgreSQL. The MADlib exploits user-defined aggregates (UDAs), user-defined functions (UDFs), and a sparse matrix C library to provide efficient data processing on disk and in memory.

The Bismarck system [5] exploits UDFs as a convenient interface for in-DBMS data mining. The Bismarck supports logistic regression, support vector machines and other mining methods, which are based on incremental gradient descent technique.

The DAnA system [13] automatically maps a high-level specification of in-DBMS analytic queries to the FPGA accelerator. The accelerator implementation is generated from an UDF, expressed as part of a SQL query in a Python-embedded Domain-Specific Language. The DAnA supports striders, a special hardware structures that directly interface with the buffer pool of the DBMS to extract and clean the data tuples, and pass the data to FPGA.

This paper extends our previous research as follows. In [23, 24], we presented an approach to encapsulation parallel mining algorithms for many-core accelerators into an open-source DBMS (with PostgreSQL as an example) for small-scale data sets. In [20, 22], we developed a method for encapsulation parallelism into an open-source DBMS (with PostgreSQL as an example). In this paper, we apply the resulting PDBMS to fuzzy clustering very large data sets, extending our previous in-PostgreSQL fuzzy clustering algorithm [16].

3 Encapsulation of Partitioned Parallelism into a Serial DBMS

3.1 Basic Ideas

Our development is based on the concept of the partitioned parallelism [4], which assumes the following. Let us consider a computer cluster as a set of alike computers (nodes) connected with high-speed network where each node is equipped with its own main memory and disk, and has an instance of PDBMS installed.

Each database table is fragmented into a set of horizontal partitions according to the number of nodes in the cluster, and partitions are distributed across nodes. The way of partitioning is defined by a fragmentation function (a table is associated with its own fragmentation function), which takes a tuple of the table as an input and returns a node where the tuple should be stored. One of the table attributes is declared as a partitioning attribute to be an argument of the fragmentation function.

One of the PDBMS instances is declared as a coordinator. A retrieve query is executed independently by all the PDBMS instances where each instance processes its own database partition and generates a partial query result, and then partial results are merged by the coordinator into the resulting table.

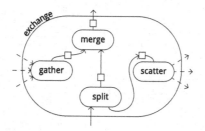

Fig. 1. Structure of the EXCHANGE operator

To implement the schema described above, PDBMS engine provides the EXCHANGE operator [28]. Such an operator is inserted into a serial query plan and encapsulates all the details related to parallelism. EXCHANGE is a composite operator (see Fig. 1) with two attributes, namely port and distribution function.

Port is serial number of the EXCHANGE operator in the query plan to provide concordance of all the query plans of PDBMS instances. The distribution function takes a tuple as an input and returns a number of the instance where the tuple should be processed.

The SPLIT operator computes the distribution function for an input tuple. If the tuple is to be processed by the current instance, it passed to the MERGE operator. Otherwise, the tuple is passed to the SCATTER operator to be sent to the respective instance. The GATHER operator receives tuples from other instances and passes them to MERGE. The MERGE operator alternately combines the tuple streams from SPLIT and GATHER.

EXCHANGE provides data transfers in case of queries where tables are joined, and partitioning attribute of the table(s) does not match to join attribute. Also, EXCHANGE with distribution function identical to the coordinator number being inserted into the root of a query plan, provides merging partial results of the query into the resulting table.

3.2 Implementation with PostgreSQL

Open-source DBMSs are now being a reliable alternative to commercial DBMSs. In this regard, an idea of obtaining a PDBMS by small-scale modifications of the original source code of an open-source serial DBMS to encapsulate partitioned parallelism looks promising. PargreSQL [20,22] is an example of PDBMS implemented on top of PostgreSQL in an above-mentioned way.

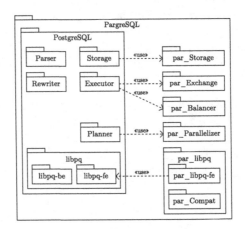

Fig. 2. Structure of PargreSQL

Figure 2 depicts module structure of PargreSQL. The original open-source DBMS is considered as one of PDBMS subsystems. Novel subsystems extend PargreSQL as follows.

The *par_Storage* subsystem provides metadata on partitioning in PDBMS dictionary. In PargreSQL, each table should contain at least one integer column to be a partitioning attribute. Thus, each CREATE TABLE command should be ended by the (WITH FRAGATTR=*pa*) clause to provide the table's fragmentation function as *pa mod P* where *pa* is partitioning attribute and *P* is the number of PDBMS instances.

The *par_Exchange* subsystem implements the EXCHANGE operator described above. The *par_Parallelizer* subsystem inserts EXCHANGEs into the appropriate places of a query plan given from PostgreSQL.

The *par_libpq* subsystem is a modified version of the PostgreSQL libpq library. Being a wrapper over the original PostgreSQL libpq-fe front-end, the *par_libpq-fe* provides connection of a client to all the PDBMS instances and replicates a query to each PargreSQL engine.

The *par_Compat* subsystem is a set of C preprocessor macros that change the original PostgreSQL API calls into the PargreSQL API calls to provide transparent migration of PostgreSQL applications to PargreSQL.

In the end, resulting parallel DBMS is obtained by modifications that took less than one per cent of whole source code of PostgreSQL.

4   Fuzzy Clustering Inside a Parallel DBMS

Clustering could be seen as a task of grouping a finite set of objects from finite-dimensional metric space in such a way that objects in the same group (called a cluster) are closer (with respect to a chosen distance function) to each other than to those in other groups (clusters). Hard clustering implies that each object must belong to a cluster or not. In fuzzy clustering, each object belongs to each cluster to a certain membership degree. Fuzzy C-means (FCM) [3] is the one of the most widely used fuzzy clustering algorithm.

Further, in Sect. 4.1, we give notations and definitions regarding FCM, and in Sect. 4.2, we show in-PDBMS implementation of FCM.

4.1   Notations and Definitions

Let $X = \{x_1, \ldots, x_n\}$ is a set of objects to be clustered where an object $x_i \in \mathbb{R}^d$. Let $k \in \mathbb{N}$ is the number of clusters where each cluster is identified by a number from 1 to k. Then $C \in \mathbb{R}^{k \times d}$ is the *matrix of centroids* where $c_j \in \mathbb{R}^d$ is center of a j-th cluster.

Let $U \in \mathbb{R}^{n \times k}$ is the *matrix of memberships* where $u_{ij} \in \mathbb{R}$ reflects membership of an object x_i to a centroid c_j and the following holds:

$$\forall\, i,j \quad u_{ij} \in [0;1], \quad \forall\, i \quad \sum_{j=1}^{k} u_{ij} = 1. \tag{1}$$

The *objective function* J_{FCM} is defined as follows:

$$J_{FCM}(X, k, m) = \sum_{i=1}^{n} \sum_{j=1}^{k} u_{ij}^m \rho^2(x_i, c_j) \tag{2}$$

where $\rho : \mathbb{R}^d \times \mathbb{R}^d \to \mathbb{R}_+ \cup 0$ is the distance function to compute proximity of an object x_i to a centroid c_j, and $m \in \mathbb{R}$ ($m > 1$) is the fuzzyfication degree of J_{FCM} (the algorithm's parameter, which is usually taken as $m = 2$). Without loss of generality, we may use the Euclidean distance, which is defined as follows:

$$\rho^2(x_i, c_j) = \sum_{\ell=1}^{d} (x_{i\ell} - c_{j\ell})^2. \tag{3}$$

FCM iteratively minimizes J_{FCM} according to the following formulas:

$$\forall j, \ell \quad c_{j\ell} = \frac{\sum\limits_{i=1}^{n} u_{ij}^m \cdot x_{i\ell}}{\sum\limits_{i=1}^{n} u_{ij}^m} \tag{4}$$

$$u_{ij} = \sum_{t=1}^{k} \left(\frac{\rho(x_i, c_j)}{\rho(x_i, c_t)} \right)^{\frac{2}{1-m}}. \tag{5}$$

Finally, Algorithm 1 depicts basic FCM.

Alg. 1. FCM(IN X, m, ε, k; OUT U)

1: $U^{(0)} \leftarrow random(0..1)$; $s \leftarrow 0$
2: **repeat**
3:     Compute $C^{(s)}$ by (4)
4:     Compute $U^{(s)}$ and $U^{(s+1)}$ by (5)
5:     $s \leftarrow s + 1$
6: **until** $\max_{ij} \{|u_{ij}^{(s+1)} - u_{ij}^{(s)}|\} \geq \varepsilon$
7: **return** U

4.2   The PgFCM Algorithm

The pgFCM algorithm is an in-PDBMS implementation of FCM. Our algorithm provides a database partitioned among the disks of computer cluster nodes, and performs fuzzy clustering by SQL queries where each database partition is processed independently by the respective instance of the PargreSQL DBMS described above.

Table 1 depicts design of pgFCM database. Underlined column name specifies primary key, and double underlined column name specifies partitioning attribute of the respective table. To improve efficiency of query execution, we provide index

Table 1. Database scheme of the pgFCM algorithm

Table	Columns	Indexed column(s)	Meaning
SH	$\underline{i}, x_1, x_2, \ldots, x_d$	i	Set of objects X, horizontal representation
SV	$\underline{i}, \ell, val$	i, ℓ, (i, ℓ)	Set of objects X, vertical representation
C	$\underline{j}, \ell, val$	ℓ, (j, ℓ)	Matrix of centroids C, vertical representation
SD	$\underline{i, j}, dist$	i, (i, j)	Distances between objects x_i and centroids c_j
U	$\underline{i, j}, val$	i, (i, j)	Matrix of memberships U at the s-th step, vertical representation
UT	$\underline{i, j}, val$	(i, j)	Matrix of memberships U at the $(s + 1)$-th step, vertical representation

file where indices by the primary key of the tables are automatically created, and the rest indices are created manually.

FCM computations (2)–(5) require aggregations over columns of the input table, which are directly impossible in SQL. To overcome this, in addition to horizontal representation of the algorithm's key data, namely set of objects and the matrix of memberships, we provide their vertical representation. Such a technique allows for using SQL aggregation functions SUM and MAX while implementing computations in FCM as a set of SQL queries.

Alg. 2. PGFCM(IN table SH, m, ε, k; OUT table U)

1: Initialize table U, table SV
2: **repeat**
3:     Compute centroids by modifying table C
4:     Compute distances by modifying table SD
5:     Compute memberships by modifying table UT
6:     **TRUNCATE U; INSERT INTO U SELECT * FROM UT**
7:     $\delta \leftarrow$ **SELECT max(abs(UT**.val **- U**.val**)) FROM UT, U WHERE UT**.i=U.i **AND UT**.j=U.j
8: **until** $\delta \geq \varepsilon$
9: **SELECT * FROM U**

Algorithm 2 depicts implementation schema of pgFCM. The algorithm is implemented as an application in C language, which connects to all PargreSQL instances and performs computations by SQL queries over tables described in Table 1.

```
1  — Initialization of table SV
2  for each cnt ∈ 1..d do
3     INSERT INTO SV
4        SELECT SH.i, cnt, x_cnt FROM SH
5  — Initialization of table U
6  for each i ∈ 1..n do
7     for each j ∈ 1..k do
8        INSERT INTO U VALUES (i,j,random(0..1))
9  UPDATE U SET val=val/U1.tmp
10    FROM (SELECT i, sum(val) AS tmp FROM U GROUP BY i) AS U1
11    WHERE U1.i=U.i
```

Fig. 3. Implementation of initialization steps of pgFCM

Figure 3 shows how to form vertical representation of the SH table, and initialize the U table according to (1).

```
1  — Computing centroids by modifying table C
2  INSERT INTO C
3     SELECT R.j, SV.ℓ, sum(R.s * SV.val)/sum(R.s) AS val
4     FROM (SELECT i, j, U.val^m AS s FROM U) AS R, SV
5     WHERE R.i=SV.i
6     GROUP BY j, ℓ;
7  — Computing distances by modifying table SD
8  INSERT INTO SD
9     SELECT i, j, sqrt(sum((SV.val−C.val)^2)) AS dist
10    FROM SV, C
11    WHERE SV.ℓ=C.ℓ;
12    GROUP BY i, j;
13 — Computing memberships by modifying table UT
14 INSERT INTO UT
15    SELECT i, j, SD.dist^(2^(1−m))*SD1.den AS val
16    FROM (SELECT i, 1/sum(dist^(2^(m−1))) AS den FROM SD
17       GROUP BY i) AS SD1, SD
18    WHERE SD.i=SD1.i;
```

Fig. 4. Implementation of computing steps of pgFCM

Figure 4 depicts computational steps of the pgFCM algorithm. While modifying the UT table, we exploit the following version of (5), which is more convenient for computations in SQL:

$$u_{ij} = \rho^{\frac{2}{1-m}}(x_i, c_j) \cdot \left(\sum_{t=1}^{k} \rho^{\frac{2}{m-1}}(x_i, c_t) \right)^{-1}. \tag{6}$$

5 The Experiments

We evaluated the proposed algorithm in experiments conducted on the Tornado SUSU supercomputer [11]. In the experiments, we compared the performance of pgFCM with the following analogs.

In [6], Ghadiri *et al.* presented BigFCM, the MapReduce-based fuzzy clustering algorithm. BigFCM was evaluated on a computer cluster of one master and 8 slave nodes over the HIGGS dataset [2] consisting of $1.1 \cdot 10^7$ the 29-dimensional objects. Hidri *et al.* [9] proposed the parallel WCFC (Weighted Consensus Fuzzy Clustering) algorithm. In the experiments, WCFC was evaluated on a computer cluster of 20 nodes over the KDD99 dataset [1] consisting of $4.9 \cdot 10^6$ the 41-dimensional objects. According to the experimental evaluation performed by the authors of the above-mentioned algorithms, both BigFCM and WCFC overtake other out-of-DBMS parallel implementations of FCM, namely MR-FCM [12] and Mahout FKM [30].

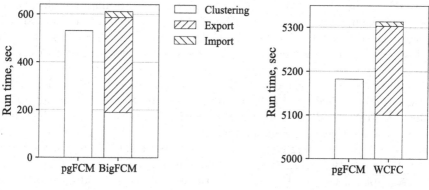

(a) pgFCM vs BigFCM over the HIGGS dataset ($n = 1.1 \cdot 10^7$, $d = 29$, $k = 2$)

(b) pgFCM vs WCFC over the KDD99 dataset ($n = 4.9 \cdot 10^6$, $m = 41$, $k = 2$)

Fig. 5. Comparison of pgFCM with analogs

We ran pgFCM on Tornado SUSU with a reduced number of nodes to make the peak performance of our system approximately equal to that of the system on which the corresponding competitor was evaluated. Throughout the experiments, we used the same datasets that were employed for the evaluation of the competitors. For comparison purposes, we used the run times reported in the respective papers [6,9].

For out-of-DBMS analogs, we measured the run time needed to export the initial dataset from the PostgreSQL DBMS (as conversion of a table to CSV file) and import the clustering results back into PostgreSQL (as loading of a CSV file into DBMS as a table), and added these overhead costs to the clustering running time of analogs. We assume the typical scenario when the data to be clustered are stored in a DBMS, and it is necessary to export the data outside the DBMS before clustering and import clustering results back into the DBMS after clustering.

Experimental results are depicted in Fig. 5. As can be seen, pgFCM is inferior to analogs in the performance of clustering. However, unlike the analogs, pgFCM performs clustering inside a PDBMS and does not need to export data and import results, and we can see that the proposed algorithm outruns analogs with respect to the overhead costs on export-import data.

6   Conclusions

In this paper, we addressed the task of mining in very large data sets inside a relational DBMSs, which remain the basic tool for data processing in various subject domains. Integration of data mining methods with relational DBMS avoids export-import bottleneck and provides the end-user with all the built-in DBMS services (query optimization, data consistency and security, etc.).

To effectively process very large databases, we encapsulate parallelism into the PostgreSQL open-source DBMS. Resulting parallel DBMS (called PargreSQL) is obtained by small-scale modifications of the original source code of PostgreSQL. Such modifications took less than one per cent of whole source code of PostgreSQL.

In this study, we implement Fuzzy C-Means clustering algorithm inside PargreSQL. The algorithms is implemented as an application in C language, which utilizes PargreSQL API. We design the algorithm's database in a way that allows for employing SQL row aggregation functions. We carry out experiments on computer cluster system using referenced data sets and compare our approach with parallel out-of-DBMS solutions. The experimental evaluation shows that the proposed approach overtakes analogs with respect to overhead on export data outside a DBMS and import results of analysis back into a DBMS.

In further studies, we plan to apply PargreSQL to other data mining problems over very large databases, e.g. association rules and classification.

Acknowledgments. The study was financially supported by the Ministry of Science and Higher Education of the Russian Federation within the framework of the Russian Federal Program for the Development of Russian Science and Technology from 2014 to 2020; project identifier: RFMEFI57818X0265 (contract no. 075-15-2019-1339 (14.578.21.0265)).

References

1. KDD Cup 1999 Data. http://kdd.ics.uci.edu/databases/kddcup99/kddcup99. html. Accessed 01 July 2019

2. Baldi, P., Sadowski, P., Whiteson, D.: Searching for exotic particles in high-energy physics with deep learning. Nat. Commun. **4**, 4308 (2014). https://doi.org/10.1038/ncomms5308

3. Bezdek, J.C.: Pattern Recognition with Fuzzy Objective Function Algorithms. Springer, New York (1981). https://doi.org/10.1007/978-1-4757-0450-1

4. DeWitt, D.J., Gray, J.: Parallel database systems: the future of high performance database systems. Commun. ACM **35**(6), 85–98 (1992). https://doi.org/10.1145/129888.129894

5. Feng, X., Kumar, A., Recht, B., Ré, C.: Towards a unified architecture for in-RDBMS analytics. In: Proceedings of the ACM SIGMOD International Conference on Management of Data, SIGMOD 2012, Scottsdale, AZ, USA, 20–24 May 2012, pp. 325–336 (2012). https://doi.org/10.1145/2213836.2213874

6. Ghadiri, N., Ghaffari, M., Nikbakht, M.A.: BigFCM: fast, precise and scalable FCM on hadoop. Future Gener. Comput. Syst. **77**, 29–39 (2017). https://doi.org/10.1016/j.future.2017.06.010

7. Han, J., et al.: DBMiner: a system for mining knowledge in large relational databases. In: Proceedings of the 2nd International Conference on Knowledge Discovery and Data Mining (KDD 1996), pp. 250–255, Portland (1996). http://www.aaai.org/Library/KDD/1996/kdd96-041.php

8. Hellerstein, J.M., et al.: The MADlib analytics library or MAD skills, the SQL. PVLDB **5**(12), 1700–1711 (2012). https://doi.org/10.14778/2367502.2367510

9. Hidri, M.S., Zoghlami, M.A., Ayed, R.B.: Speeding up the large-scale consensus fuzzy clustering for handling big data. Fuzzy Sets Syst. **348**, 50–74 (2018). https://doi.org/10.1016/j.fss.2017.11.003

10. Imielinski, T., Virmani, A.: MSQL: a query language for database mining. Data Min. Knowl. Discov. **3**(4), 373–408 (1999). https://doi.org/10.1023/A:1009816913055

11. Kostenetskiy, P., Semenikhina, P.: SUSU supercomputer resources for industry and fundamental science. In: 2018 Global Smart Industry Conference (GloSIC), Chelyabinsk, Russia, 13–15 November 2018, p. 8570068 (2018). https://doi.org/10.1109/GloSIC.2018.8570068

12. Ludwig, S.A.: MapReduce-based fuzzy c-means clustering algorithm: implementation and scalability. Int. J. Mach. Learn. Cybern. **6**(6), 923–934 (2015). https://doi.org/10.1007/s13042-015-0367-0

13. Mahajan, D., Kim, J.K., Sacks, J., Ardalan, A., Kumar, A., Esmaeilzadeh, H.: In-RDBMS hardware acceleration of advanced analytics. PVLDB **11**(11), 1317–1331 (2018). https://doi.org/10.14778/3236187.3236188

14. McCaffrey, J.D.: A hybrid system for analyzing very large graphs. In: 9th International Conference on Information Technology: New Generations, ITNG 2012, Las Vegas, Nevada, USA, 16–18 April 2012, pp. 253–257 (2012). https://doi.org/10.1109/ITNG.2012.43

15. Meo, R., Psaila, G., Ceri, S.: A new SQL-like operator for mining association rules. In: Proceedings of 22th International Conference on Very Large Data Bases, VLDB 1996, 3–6 September 1996, Mumbai, India, pp. 122–133 (1996). http://www.vldb.org/conf/1996/P122.PDF

16. Miniakhmetov, R., Zymbler, M.: Integration of the fuzzy c-means algorithm into PostgreSQL. Numer. Methods Program. **13**, 46–52 (2012). https://num-meth.srcc.msu.ru/english/zhurnal/tom_2012/v13r207.html

17. Ordonez, C.: Integrating k-means clustering with a relational DBMS using SQL. IEEE Trans. Knowl. Data Eng. **18**(2), 188–201 (2006) https://doi.org/10.1109/TKDE.2006.31

18. Ordonez, C.: Can we analyze big data inside a DBMS? In: Proceedings of the 16th International Workshop on Data warehousing and OLAP, DOLAP 2013, San Francisco, CA, USA, 28 October 2013, pp. 85–92 (2013). https://doi.org/10.1145/2513190.2513198

19. Ordonez, C., Pitchaimalai, S.K.: Bayesian classifiers programmed in SQL. IEEE Trans. Knowl. Data Eng. **22**(1), 139–144 (2010). https://doi.org/10.1109/TKDE.2009.127

20. Pan, C.S., Zymbler, M.L.: Taming elephants, or how to embed parallelism into PostgreSQL. In: Decker, H., Lhotská, L., Link, S., Basl, J., Tjoa, A.M. (eds.) DEXA 2013. LNCS, vol. 8055, pp. 153–164. Springer, Heidelberg (2013). https://doi.org/10.1007/978-3-642-40285-2_15

21. Pan, C.S., Zymbler, M.L.: Very large graph partitioning by means of parallel DBMS. In: Catania, B., Guerrini, G., Pokorný, J. (eds.) ADBIS 2013. LNCS, vol. 8133, pp. 388–399. Springer, Heidelberg (2013). https://doi.org/10.1007/978-3-642-40683-6_29

22. Pan, C.S., Zymbler, M.L.: Encapsulation of partitioned parallelism into open-source database management systems. Program. Comput. Softw. **41**(6), 350–360 (2015). https://doi.org/10.1134/S0361768815060067

23. Rechkalov, T., Zymbler, M.: An approach to data mining inside PostgreSQL based on parallel implementation of UDFs. In: Selected Papers of the XIX International Conference on Data Analytics and Management in Data Intensive Domains (DAMDID/RCDL 2017), Moscow, Russia, 9–13 October 2017, vol. 2022, pp. 114–121 (2017). http://ceur-ws.org/Vol-2022/paper20.pdf

24. Rechkalov, T., Zymbler, M.: Integrating DBMS and parallel data mining algorithms for modern many-core processors. In: Kalinichenko, L., Manolopoulos, Y., Malkov, O., Skvortsov, N., Stupnikov, S., Sukhomlin, V. (eds.) DAMDID/RCDL 2017. CCIS, vol. 822, pp. 230–245. Springer, Cham (2018). https://doi.org/10.1007/978-3-319-96553-6_17

25. Sattler, K., Dunemann, O.: SQL database primitives for decision tree classifiers. In: Proceedings of the 2001 ACM CIKM International Conference on Information and Knowledge Management, Atlanta, Georgia, USA, 5–10 November 2001, pp. 379–386 (2001). https://doi.org/10.1145/502585.502650

26. Shang, X., Sattler, K.-U., Geist, I.: SQL based frequent pattern mining with FP-growth. In: Seipel, D., Hanus, M., Geske, U., Bartenstein, O. (eds.) INAP/WLP -2004. LNCS (LNAI), vol. 3392, pp. 32–46. Springer, Heidelberg (2005). https://doi.org/10.1007/11415763_3

27. Sidló, C.I., Lukács, A.: Shaping SQL-based frequent pattern mining algorithms. In: Bonchi, F., Boulicaut, J.-F. (eds.) KDID 2005. LNCS, vol. 3933, pp. 188–201. Springer, Heidelberg (2006). https://doi.org/10.1007/11733492_11

28. Sokolinsky, L.B.: Organization of parallel query processing in multiprocessor database machines with hierarchical architecture. Program. Comput. Softw. **27**(6), 297–308 (2001). https://doi.org/10.1023/A:1012706401123

29. Sun, P., Huang, Y., Zhang, C.: Cluster-by: an efficient clustering operator in emergency management database systems. In: Gao, Y., et al. (eds.) WAIM 2013. LNCS, vol. 7901, pp. 152–164. Springer, Heidelberg (2013). https://doi.org/10.1007/978-3-642-39527-7_17

30. Xhafa, F., Bogza, A., Caballé, S., Barolli, L.: Apache Mahout's k-Means vs Fuzzy k-Means performance evaluation. In: 2016 International Conference on Intelligent Networking and Collaborative Systems, INCoS 2016, Ostrawva, Czech Republic, 7–9 September 2016, pp. 110–116 (2016). https://doi.org/10.1109/INCoS.2016.103
31. Zymbler, M., Kumar, S., Kraeva, Y., Grents, A., Perkova, A.: Big data processing and analytics inside DBMS. In: Selected Papers of the XXI International Conference on Data Analytics and Management in Data Intensive Domains (DAMDID/RCDL 2019), Kazan, Russia, 15–18 October 2019, p. 21 (2019). http://ceurws.org/Vol-2523/invited04.pdf

Data Science for Education

Issues and Lessons Learned in the Development of Academic Study Programs in Data Science

Ivan Luković[(✉)] [iD]

Faculty of Technical Sciences, University of Novi Sad, Novi Sad, Serbia
ivan@uns.ac.rs

Abstract. In recent years, Data Science has become an emerging education and research discipline all over the world. Software industry shows an increasing and even quite intensive interest for academic education in this area. In this paper, we announce main motivation factors for creating a new study program in Data Science at Faculty of Technical Sciences of University of Novi Sad, and discuss why it is important to nurture the culture of interdisciplinary orientation of such program from early beginning of undergraduate studies. Also, we announce how we structured the new study program and addressed the main issues that come from evident industry requirements. The program was initiated in 2017, both B.Sc. and M.Sc. level, and we present in the paper the experiences collected through the first three years of its execution.

Keywords: Academic education · Data Science · Information Engineering

1 Introduction

In 2015, the three study programs in Data Science were accredited at the University of Novi Sad, Faculty of Technical Sciences (FTS) in Novi Sad. One is a 4-year B.Sc. program in Information Engineering, and the two are master-level study programs: a) 1-year M.Sc. in Information Engineering, and b) 1,5-year M.Sc. in Information and Analytics Engineering. All the programs are officially accredited in the category of interdisciplinary and multidisciplinary programs, in the two main areas by the official classification of education areas in Republic of Serbia: Electrical Engineering and Computing, and Engineering Management. In practice, the programs cover in deep the disciplines in Data Science, as a completely new combination of courses predominantly coming from Computer Science, Software Engineering, Mathematics, Telecommunications and Signals, Finances, and Engineering Management.

Execution of the two of these study programs has been initiated in 2017. Those are B.Sc. in Information Engineering, and M.Sc. in Information Engineering. By this, now we have three active generations of students at both levels. Our first experiences with these generations of students are quite positive.

Design of the aforementioned study programs were motivated mainly by the idea to profile specific study programs in the scope of Computer Science, Informatics, or Software Engineering (CSI&SE) disciplines, so as to nurture the appropriate level of interdisciplinarity and contribute to resolving the following two paradoxes:

© Springer Nature Switzerland AG 2020
A. Elizarov et al. (Eds.): DAMDID/RCDL 2019, CCIS 1223, pp. 227–245, 2020.
https://doi.org/10.1007/978-3-030-51913-1_15

(P1) More interdisciplinary oriented experts, capable of covering a wide range of tasks, knowledge and skills are always significantly better positioned in the software industry Human Resource (HR) market, while academic institutions offer study programs that are rather self-contained, i.e. oriented to a narrower knowledge scope.

(P2) Students or young software engineers believe that they will be better positioned in software industry HR market just as they are good IT experts, i.e. programmers, while employers rather expect experts capable of recognizing and resolving their interdisciplinary oriented and complex requirements.

Our goal was to create such structure and content of the new study programs to address the identified paradoxes and main issues that come from evident industry requirements.

In our current academic education, we can identify study programs of the three categories, covering in some extent disciplines of CSI&SE, as a basis to provide Data Science education. Those are: (1) Specific study programs in CSI&SE; (2) Study programs in (Applied) Mathematics; and (3) Study programs in Economics, Business Administration and Management. Our experiences in teaching CSI&SE courses in study programs of all the three categories lead to the identification of typical students' and even teachers' behavioral patterns. In the paper we will discuss why such patterns lower the culture of interdisciplinarity, and how it can be raised by Data Science study programs. Also, we communicate in the paper some our recent experiences from the execution of the Information Engineering study programs, where we identify increasing awareness of students about the importance of Data Science in upcoming years, while still we notify a polarization of students' population to one, with clear ideas about their future, vs. students with not clear recognition of their future opportunities.

The paper is structured as follows. In Related Work we further discuss main motivation factors for reconceptualization or renovation in higher education that may be implemented by creating a new study program in Data Science, and advocate why it is important to nurture the culture of interdisciplinary orientation of such program. In Sect. 3 we advocate that strong interdisciplinary and multidisciplinary nature of study programs is a key requirement in Data Science academic education. Section 4 gives a short description of the new study programs in Data Science, accredited at FTS. In Sect. 5 we communicate our experiences related to the accreditation process of our study programs in 2015, their execution from 2017, and also reaccreditation process in 2019.

This paper is an extension of the abstract, published in [7].

2   Related Work

Nowadays, modern business include acquisition and store of enormous data volumes, even larger than ever before. A volume of collected data shows practically exponential growth all over the world [11, 12]. In [2], Chiang, et al. state that Big Data is an emerging phenomenon. Computing systems today are generating 15 petabytes of new information every day—eight times more than the combined information in all the libraries in the U.S.; about 80% of the data generated everyday is textual and unstructured data.

Most often, collected data are used in a shorter time frame, and then they are archived and almost not used, effectively. On the other hand side, such data represent a significant value that a company can utilize so as to reach created goals and provide a sustainable development [8]. In [14], Tulasi state that higher education is a field where tremendous amounts of data are available, while many institutions fail to make efficient use of the huge amount of data available. On the other hand side, data analytics over big data sets would lead to greater benefits and achievements. According to the McKinsey Global Institute Report [10], Big Data is the next frontier for innovation, competition, and productivity.

Typically, a company management is aware about significant but not exploited values ingrained in stored high data volumes. Also, we believe that the needs for generation of company knowledge form data are clearly recognized in well-matured companies. Such knowledge is to be used to raise the effectiveness of the decision and management company processes, based predominantly on quantitative, analytical methods [8, 9].

As we discussed in [8], unfortunately, a daily practice in many companies still intensively points out to the problem of a serious gap between the identified needs for knowledge, on one hand side, and inability of modern software products to address such needs in an effective way, on the other hand side, despite that massive data volumes already exist, while modern Information Technologies (ITs) provide the excellent technology prerequisites for a development and industry implementation of high quality software applications. We believe that this problem is just a new, "modern" form of never-ended software crisis present for decades in many different forms, in software industry. We can even call such phenomenon a "data crisis". Some of important causes of the aforementioned phenomenon are in the following:

1. Unsatisfactory level of organization maturity in regard to the: capacities for information management, quality management, and business processes;
2. Unsatisfactory level of accumulated knowledge in a problem domain; and
3. Unsatisfactory level of accumulated knowledge in a domain of software engineering, particularly in a domain of the development and formal specification of models for software products aimed at generation of company knowledge and decision support.

Alleviating the aforementioned problem is a strategic and long life task, only possible by simultaneous addressing all its significant causes. By this, addressing the cause (3) is an important endeavor in regard to the aspect of formal academic education. Following the aforementioned paradoxes and phenomenon, in well-developed countries a highly emerging interest for studying a wide range of knowledge in data analytics, big data processing and generation of company knowledge is present in recent years [1, 4, 5]. Such interest leads to the discipline of Data Science [12, 13]. We notify a more intensive interest in well-developed countries for Data Science academic education even in years from 2015. The similar is also notified in [16], where much more study programs are developed at graduate, then at undergraduate level.

In [2], the authors advocate that Business Intelligence and Analytics (BI&A), i.e. Data Science applied to business, has gained much attention in both the academic and IT practitioner communities over the past two decades. They state that universities are beginning to respond to the research and educational demands from industry and discuss the role of Information Systems (IS) curricula in delivering BI&A education, the challenges facing IS departments, and the new vision for the IS discipline. They identify the skills necessary for BI&A, and classify them as: (i) analytical, (ii) IT knowledge, and (iii) Business knowledge and communication skills.

Nowadays, a predominant attitude is that the knowledge required for development and application of Data Science is highly interdisciplinary and multidisciplinary oriented, as there is a strong need to apply various knowledge and disciplines in a unified way in addressing the complex business problems, as a key issue in modern business [8]. In [15], Wixom, et al. present their findings that universities can produce students with a broader range of Business Intelligence (BI) skills using an interdisciplinary approach within BI classes and programs; and academic BI offerings should better align with the needs of practice.

According to [3], there exist wide range of critical issues that need to be considered when working with Big Data in education. Daniel in [3] identifies the issues that include diversity in the conception and meaning of Big Data in education, ontological, epistemological disparity, technical challenges, ethics and privacy, digital divide and digital dividend, lack of expertise and academic development opportunities to prepare educational researchers to leverage opportunities afforded by Big Data. The author proposes Data Science as the fourth research methodology tradition in educational research. He advocates for a reconceptualization of value and relevance of Big Data in educational research, and creating educational research programs in Data Science that support the successful implementation of Big Data in education.

Klašnja-Milićević, et al. in [6] summarize a significance of coupling Big Data and Learning Analytics in Data Science education, and propose a widespread architecture framework that could be achieved by exhausting big data techniques in the field of education. They highlight the significance of big data and analytics in education, which is twofold: in managing reform activities in higher education, and assisting instructors in improving teaching and learning. In [14], Tulasi advocates that Big Data and Analytics will play a significant role in future of higher education. By this, analytics would lead to innovation in education, which can be of two types: sustainable and disruptive. The first is to motivate continuous improvements of the existing systems and processes, while the former is to introduce new ideas and activities that will significantly break the current practice in creating new systems. By Tulasi, both forms of innovation are required to adapt to the growing needs in higher education. In [11], Moreira, et al. state that general behavioral changing of societies and younger generations has a great influence in the way the young people perceive higher education, which requires a disruption of current teaching-learning models. The authors advocate inclusion in this process the technology and habits of the daily lives of the generations that are coming to higher education. They propose a disruptive conceptual approach to the teaching-learning process and extend it with the Learning Analytics component.

In recent years, various Data Science study programs have been introduced at many world-wide universities. Their content and structure may significantly differ from institution to institution, due to: different nature, study approaches and rules being applied; various levels of expertise; and coverage of various disciplines that may include but not to be constrained to computer science, mathematics and statistics, system engineering, business and management, and selected problem domain topics. Yan and Davis in [16] notify that the University of Massachusetts Dartmouth from 2015 offers degree programs in Data Science, at both the undergraduate and graduate levels. After their finding that just few articles have been published that deal with graduate Data Science courses, and even much less dealing with undergraduate ones, they focus on the structure and courses at undergraduate level. Besides, they propose a structure and the content of an introductory course in Data Science. The authors identify the notion of the *data science life cycle* and base a design of their introductory course around the concepts related to this notion.

Following all these references, as well as many others, we believe that academic education in Data Science, and particularly education of engineers of such a profile, is also a strong issue in upcoming years, and we have invested our efforts in addressing it at the University of Novi Sad, Faculty of Technical Sciences in Novi Sad by creating the new undergraduate and graduate study programs in Information Engineering. Our intention was to properly: (i) address strong requirements regarding the level of their interdisciplinary and multidisciplinary nature; (ii) achieve a good balance between continuous changes and improvements vs. disruptive actions and improvements; and (iii) pay a particular attention to the design of introductory undergraduate courses, keeping in mind their high sensitivity and great influence to the overall success of the undergraduate study program in Data Science.

3 Interdisciplinarity as a Key Requirement in Data Science Academic Education

As we already discussed in [8], in software industry of well-developed economies, we can notify a strong fitting between the skill and education requirements for specific job positions, and the level of education and experience of software engineers or IT experts being hired at those positions. Also, such software engineers typically show a higher level of specialization to some disciplines or problem domains, as HR market is more mature. On the contrary, in under-developed or even developing economies, HR market in software industry is not as mature as in well-developed economies. Therefore, fitting between the required skills and education level for the job positions and the level of education and experience of software engineers is often not appropriate, and we can notice hiring overqualified or underqualified experts at some positions, in a wider extent. The level of specialization depth of software engineers to some disciplines or problem domains is not as strong, as in well-developed economies. However, nevertheless if we observe a software industry in well-developed or under-developed economy, the paradoxes (P1) and (P2), given in Sect. 1 arise. If we say here that academic institutions motivate, often "in silence", education of "more specialized experts", that means experts that are pure software or informatics engineers, mathematicians, business administration managers, or various domain experts.

On the contrary, interdisciplinary and multidisciplinary characteristics and skills are to be nurtured from early ages of academic education. Otherwise, by our previous experiences, we are just the witnesses of an evident and almost remediless attrition of such skills and characteristics in the student population.

We identify study programs of the three categories, covering in some extent disciplines of CSI&SE, as a basis to provide Data Science education. Those are: (1) Specific study programs in CSI&SE; (2) Study programs in (Applied) Mathematics; and (3) Study programs in Economics, Business Administration and Management. All large Serbian universities provide study programs from all the three categories, for many years, as it is the similar in many other countries. Typical students' behavioral patterns of all the three categories [8, 9] are as follows.

(1) Students from specific study programs in CSI&SE are predominantly technology oriented. Often, they express their animosity to the mathematical, and even more organizational, managerial or economics disciplines, with a belief that this knowledge is not necessary to them, and that someone else is to posses it. Study programs of this category often provide just a modest level of knowledge from mathematics and business administration. On the other hand side, such students express their strong interest for learning a typical technology knowledge in IT. By this, we name this behavioral pattern as "Let me learn one more technology environment, only".

(2) Students from study programs in Applied Mathematics, or just Mathematics are predominantly formally oriented. They believe that technology knowledge is of a lower level value, and also they are not aware of a necessity of having a knowledge from business administration, management, or economics. Development of skills aimed at practical application of adopted knowledge in various application domains is often underestimated or even neglected. Students from this category believe that complexity of things is just of a logical nature – the things are more complex, just if they are logically complex, while other forms of complexity are rather neglected. Study programs of this category often provide a modest level of CSI&SE knowledge, as well as business administration knowledge. By this, we name this behavioral pattern as "Let me prove one more theorem, only".

(3) Students from study programs in Economics, Business Administration and Management show a strong awareness about the importance of having the CSI&SE and Mathematics knowledge in resolving the complex problems in organization systems. However, in a lack of their formal knowledge from these disciplines, they believe that someone else is to resolve such problems, while their task is just to rent high quality CSI&SE and Mathematics experts to resolve the problems. Study programs from this category motivate learning a highly formalized knowledge from CSI&SE and Mathematics rarely. By this, we name this behavioral pattern as "Let me follow the things globally and rent experts for strong and formal details".

Literally, we may say that the three identified behavioral patterns form "a universe of not joinable worlds" [8]. As such, a question arises:

(Q1) Whether such, traditional approaches to teaching selected Data Science topics can produce the appropriate experts capable of resolving complex engineering problems by the utilization of big data volumes and technologies, as well as formal modeling and data analytics methods?

A derived question is:

(Q2) Who is capable of creating study programs that will successfully provide interfaces between all required disciplines in resolving complex organizational problems by a support of software systems?

Our approach to addressing these questions is disruptive: to profile completely new, specific study programs in the scope of CSI&SE that will nurture the appropriate level of interdisciplinarity and contribute to resolving the (P1) and (P2) paradoxes.

So as to test the hypothesis about the importance of creating specific study programs in Data Science, in November 2016 we performed a short survey of the current state and needs of software companies in Serbia for the knowledge and experts in this discipline [9]. We examined, in what extent it is already present or will be present in Serbia. Results of the survey were presented in more details in [8], while we communicate here just a short summary. We have found that a significant number of IT companies in Serbia, i.e. more than three quarters of all the companies, identify their strong requirements for completing Data Science projects and hiring Data Science engineers. As expected, the number of required Data Science engineers is not as huge as for general purpose software engineers. However, many companies declared their plans to hire Data Science engineers at the level of more than 5, and even up to 50 such experts. Majority of respondents notified a lack of available Data Science engineers in the HR market, and evident difficulties in acquiring the required number of such experts. They believe that current study programs in Serbia do not produce Data Science engineers having all the required knowledge, and recommend creating specific study programs in Data Science, as the needs for such expert will significantly emerge in a near future, while the discipline of Data Science is seen as one of the most promising in IT sector. Since the needs for Data Science engineers are not as huge as for general purpose software engineers, study programs in this discipline are not to be massive in terms of the maximal number of allowed students, i.e. they should be designed for a smaller number of students.

Following our previous teaching experiences, modern trends in many recognized world-wide universities, as well as the results obtained from the survey of software industry needs for Data Science experts, already presented in [8], we believe that the answer to the Q1 question is that the approaches to teaching selected Data Science topics through traditional CSI&SE, Mathematics, Economics, Business Administration and Management study programs are not enough to address the aforementioned paradoxes. As an answer to the Q2 question, we believe that profiling specific Data Science study programs in the scope of CSI&SE can successfully address the (P1) and (P2) paradoxes.

4 Study Programs in Information Engineering

As we presented in [8], our strategic decision for creating study programs in Data Science at FTS was to cover both undergraduate (B.Sc.) and graduate (M.Sc.) academic levels. Also, we decided to design a study program for a maximum of 60 students, as the intention was not to create a massive study program. The main goals were to create a curriculum covering a body of knowledge of Data Science and Information Engineering, necessary to support information management in organization systems. The curriculum should provide a body of knowledge applicable in a wide variety of organizations of all types, or even in a domain of the research, research applications, and scientific computations. It should cover a wide range of aspects of information management, typically required by many stakeholders. It is supposed to nurture both interdisciplinary and formal approaches, where typical expected formality is at the level of mathematical rigor, whenever is possible. As Data Science is highly application oriented discipline, we decided to provide Financial Engineering, as the first application domain. Also, we left the room open for further development of various application domains.

Faculty of Technical Science from University of Novi Sad is the largest polytechnic – engineering school in Serbia and wider region, with more than 14000 active students. Education process is organized through 13 departments, offering about 100 study programs at all education levels, including academic and professional studies at undergraduate, graduate, and doctoral programs, with a coverage of almost all traditional and "modern" engineering disciplines. By this, it was a "natural place" for introducing the new study programs in Data Science, as we had all necessary capacities, before all in terms of human resources, to meet the required level of interdisciplinary and multidisciplinary nature of the newly created programs.

All the three programs, i.e. 4-year B.Sc. program in Information Engineering, and the two graduate programs, 1-year M.Sc. in Information Engineering, and 1,5-year M.Sc. in Information and Analytics Engineering, are officially accredited in the category of interdisciplinary and multidisciplinary programs, in the two main areas: Electrical Engineering and Computing, and Engineering Management. In practice, the programs cover in deep the disciplines in Data Science, as a completely new combination of courses predominantly coming from Computer Science, Software Engineering, Mathematics, Computer Engineering, Telecommunications and Signals, Finances, and Engineering Management. In this way, Department of Computing and Control is a home department of all these study programs, with a strong participation of the following faculty departments: Department of Power, Electronics, and Telecommunication Engineering; Department of Industrial Engineering and Engineering Management, Department of Fundamental Disciplines in Engineering (including Mathematics); and Department of Technical Mechanics.

We designed the Information Engineering study programs in a way to satisfy the following main didactic principles: a) Abstraction and Formalization, b) Quantification and Metrics, c) Specification and Implementation, and d) Communication skills. Abstraction and Formalization skills develop students' ability to understand and formalize application domain knowledge, problems and requirements, as well as to create meta-models, languages, concepts, or any kind of formalisms necessary to provide modeling of any knowledge in systems being observed. Quantification and Metrics skills develop students' ability to quantify, measure, analyze, simulate, and optimize anything that is required in any business or research, by comprehensive methods. Specification and Implementation skills develop ability to efficiently specify, develop, implement, and apply any software to address various information management requirements in business or research. Communication skills develop ability to successfully communicate and negotiate with other professionals, having different levels and range of knowledge.

Our intention is not to present here the whole study program structure, with a complete list of all designed courses. However, first we point out to the main disciplines, covered in a larger or smaller extent by these study programs. Those are the following disciplines:

Computer Science, Informatics, and Software Engineering, covering in detail all core CSI&SE disciplines according to the ACM and IEEE Computer Science curricula, including Programming, Programming Languages, Computer Architecture, Operating Systems, Compilers, Databases, and also including Algorithms, Formal Methods, Computational Intelligence and Machine Learning, Human-Computer Interfaces, Software Engineering, and Information Systems.

Applied Mathematics, including traditional disciplines for engineers, such as Calculus, Advanced Calculus, Algebra, Numerical Calculations, and Probability Theory, as well as modern disciplines, such as Discrete Mathematics, Combinatorics, Logic, Graph Theory, Statistics, Operational Research, and Optimization Methods.

Economics, Communicology, and Management, covering the basics of Finances and Financial Engineering, Entrepreneurship in IT sector, Risks Management, Investments Risks, Decision Theory, Business Intelligence, and Communicology, with the elements of Industrial Psychology.

General Engineering Disciplines, such us Mechanics, Time Series Processing, and Information Theory.

Apart from mandatory courses covering those disciplines in the first five semesters of the B.Sc. program in Information Engineering, students can profile themselves better according to their affinities, by selecting some of elective courses in 3rd and 4th year, organized in the two main tracks: Analytical Engineering and Applied Information Engineering. The Analytical Engineering track is more oriented to traditional Computer Science and Operational Research topics, while Applied Information Engineering is more oriented to the disciplines of Management and Economics. Still, both tracks share five courses in common.

One of the viewpoint onto the B.Sc. program in Information Engineering is by chains of related courses belonging to the aforementioned disciplines, and we present it here. In the following text, notation *Course_A* → *Course_B* denotes that some course (or alternatively a set of courses) *Course_A* in a study program precedes a course *Course_B* (or alternatively a set of courses), and is a prerequisite for it. Notation (M) next to the course name denotes that a course is mandatory in the study program, while (E) denotes that it is elective one.

In *Computer Science, Informatics, and Software Engineering* discipline, the study program provides a chain of courses covering programming methods, techniques, and technologies: Fundamentals of Programming and Programming Languages (M) → Theory of Algorithms (M) → Advanced Programming and Programming Languages (M) → Web Programming (M) → Compilers (M) → (Internet Software Architectures (E), Theory of Algorithms and Computational Complexity (E)). It interleaves with the chain of courses in computer architecture, networks, and technologies: Computer Architecture (M) → (Operating Systems (M), Introduction to Digital Systems Design (M)) → Parallel Computing (M) → (Internet Networks (E), Computer Communications (E),

Communication Systems Design (E), Computer System Design (E)). They are followed by the chain of software engineering and information system courses: Databases 1 (M) → (Software Specification and Modeling (M), Human-Computer Interaction (E)) → Databases 2 (M) → (Advanced Information System Architectures (E), Service Oriented Architectures (E), Information System Engineering (E), Database Systems (E), Software Standardization and Quality (E), Mobile Apps (E), Software Agents (E)).

In *Applied Mathematics* discipline, the study program provides a chain of traditional and applied mathematical courses: Algebra (M) → (Mathematical Analysis 1 (M), Fundamentals of Graph Theory and Combinatorics (M)) → (Mathematical Logic (M), Probability and Stochastic Processes (M)) → (Numerical Algorithms and Numerical Software (M), Practicum in Statistics(M)). It is followed by the chains of courses in *General Engineering* and, particularly computational intelligence disciplines: (Introduction to Information and Financial Engineering (M), Mechanics (M)) → (Optimization Algorithms and Nonlinear Programming (M), Time Series Data Processing (M)) → Methods and Techniques in Data Science (M) → (Operational Research (E), Machine Learning 1 (E), Soft Computing (E), Self-Learning and Adaptive Algorithms (E), Introduction to Information Theory (E), Knowledge-Based Systems (E), Machine Learning 2 (E), Knowledge Engineering (E), Biomechanics (E), Dynamics and Optimization of Engineering Systems (E), Reliability of Technical Systems (E)).

In *Economics, Communicology, and Management* discipline, the study program provides a chain of courses: (Introduction to Information and Financial Engineering (M), Communicology (M)) → Fundamentals of Financial Engineering 1 (M) → Fundamentals of Financial Engineering 2 (M) → (Decision Making (E), Risks in Investment Management (E), Corporate Finance (E), Financing of Innovative Enterprises (E), Entrepreneurship in Information and Communication Technologies (E), Performance Indicators of the Company (E), Business Information Systems (E), Principles of Economics (E), Basics of LEAN Production (E), Service Engineering (E), Business Law (E), Principles of Engineering Management (E)).

At the M.Sc. level, a pool of more than 60 elective courses is offered, covering the disciplines of Data Science, Information Engineering, High-Performance Computing (HPC), and Financial Engineering. By this, we offer many possibilities for students to profile themselves, according to their affinities, or already having jobs. Apart from traditional M.Sc. courses that we offered in Computing and Control program, covering disciplines of Computer Engineering, Control Systems, Software Engineering, Information Systems, Intelligent Systems, E-Business Systems, and Multimedia and Computer Games, in the Information Engineering and Information and Analytics Engineering study programs we offer a variety of specialized courses from all declared disciplines. Some of them are: Data Mining and Data Analysis, Data Warehouse Systems and Business Intelligence, Modeling and Optimization by Learning from Data, Bioinformatics Algorithms, Business Process Modeling, Data Compression, Neural Networks, Deep Learning Methods, Formal Methods for Modeling Software Systems, Domain Specific Modeling and Languages, Statistics in Information Engineering, Game Theory, Algorithmic Trading, Business Case Study Solving, Quantitative Methods of Risk Management, etc. New HPC courses included in the selection are: Parallel and Distributed Architectures and Languages, Parallel and Distributed Algorithms and Data Structures, HPC Systems, Big

Data Architectures, Cloud Computing, HPC in Scientific Research, and HPC in Data Science.

5 Past and Recent Experiences

In this section, we communicate our experiences related to the design of our Data Science study programs, as well as their first accreditation process in 2015. Then, we discuss the experiences related to the enrolment process, and finally the experiences come from their execution from 2017, as a basis for the reaccreditation process performed in 2019.

5.1 Design and Accreditation of the Study Programs

Our first experiences about the new study programs in Data Science came from its initial design and the first accreditation process that was conducted from 2013 to 2015 year. The initial idea was to offer students a program that would combine computer science, mathematics and other engineering disciplines, so as to create a knowledge and skills necessary to produce (business) value from data collected in various organization systems. The notion of Data Science was even rarely used in that time. According to the national accreditation process in Serbia, for each study program it was necessary to justify a compliance to the "Standard 6 - Quality, Contemporaneity, and International Compliance of a study program". To do so, one of the requirements was to present references to at least two recognized European, and one world-wide universities and their similar academic study programs, and analyze a compliance of our programs with respect to the selected international programs. Despite that we finally found such institutions and programs, it took considerable time to search for the programs under the title of Information Engineering, or even Data Science. On the contrary, in the new reaccreditation process that we conduced from 2018 to 2019 year, the same task was almost trivial, as the number of such universities was more than significantly higher. Nevertheless, similar academic study programs still exists under quite different titles, all over the world, as the discipline of Data Science or Information Engineering is still young, despite that it comprises very traditional and often fundamental engineering and mathematical disciplines.

By this, we have referenced the following international study programs, and analyzed the compliance of our programs with them. For 4-year B.Sc. program in Information Engineering, we have referenced the programs: (i) Computer Science and Management Science (BSc Hons) from the University of Edinburgh, UK; (ii) Bachelorstudium Software & Information Engineering from the Technical University of Vienna, Austria; and (iii) Data Science and Analytics from the University of Essex, UK. For 1-year M.Sc. in Information Engineering, we have referenced the clusters of programs: (i) Data Science, Engineering in Computer Science, Computer Science, and Management Engineering from the Sapienza University of Rome, Italia; (ii) Data Science and Computational Finance from the University of Essex, UK; and (iii) MS in Financial Engineering, MS in Business Analytics, MS in Operational Research, and MS in Management Science and Engineering, from the Columbia University in the City of New York, Department of Industrial Engineering and Operations Research, USA. For 1,5-year M.Sc. in Information and Analytics Engineering, the first two referenced programs from the Sapienza

University of Rome and the University of Essex are the same as for M.Sc. in Information Engineering, while the third one is: Master "Data Engineering and Analytics" from Technical University of Munich, Department of Informatics, Germany.

As Data Science and Information Engineering are emerging disciplines, such study programs equally exist at different institutions of a mathematical, engineering, business administration, or economics and finances provenance. At each of such institutions, those study programs are profiled by the courses belonging to the fundamentals of the specific areas of interests. In our study programs, we intended to cover a diversity of those disciplines in a satisfactory way. At M.Sc. level studies, different programs typically share large pools of common elective courses, and therefore it was not always enough to consider just one study program at some international institution. Therefore, in some cases, e.g. for Sapienza University and Columbia University, we have analyzed a compliance by considering the clusters of closely related international programs.

Our strategy for a design of study programs in Data Science was based on the following decisions:

1. Create completely new programs;
2. Create new programs both at B.Sc. and M.Sc. levels; and
3. Include existing courses from other study programs instead of creating completely new courses, whenever possible.

The main reason for decision (1) was in line to the disruptive approaches, identified in [14]. It was to motivate introducing fresh ideas, reaching significantly new values, and reinforcement or improving positive organization culture in the education process, while breaking some current practice adjusted to the study programs with huge numbers of students. Decision (2) was introduced to contribute to nurturing the culture of interdisciplinary and multidisciplinary approaches from the early beginning of a study process. The main reason for decision (3) was in tailoring current positive experiences and their embedding into the new program. For many years, FTS has offered numerous and valuable courses that fit well into the discipline of Data Science or Information Engineering. However, that courses are spread through various departments and study programs, with no opportunities given to students to select them together. One of the issues in creating our new programs was also to create a good composition of already existing courses and topics. By decision (3), we have created about 25% of completely new courses, while the rest of 75% were existing courses, included from other study programs. In this way, we believed that we applied a continuous approach according to [14].

Implementation of such our strategy required significant organizational efforts. The fact that our new study programs in Information Engineering included topics and courses from five departments meant a lot of talk, negotiation and deep justification of the principal idea of interdisciplinarity, so as to be well recognized among teaching staff and faculty management. During that time, we managed to noticeably improve the level of the study programs recognition and visibility among the faculty and even university teaching staff. After successful accreditation in 2015, it took another two years for the study programs to initiate their execution. To the best of our knowledge, the main cause for this two year delay was in a low recognition of the main values reached by the created study programs by the faculty management. On the contrary, all the time from 2015 we faced with very strong recognition and pushing from software industry of Serbia to initiate the execution of the created study programs. Finally, in 2017 we initiated the first execution of the two study programs: B.Sc. in Information Engineering, and M.Sc. in Information Engineering. By this, now we have three active generations of students at both levels.

5.2 Enrollment Process and Study Program Recognition

The program B.Sc. in Information Engineering is accredited with a capacity of max. 60 students. In Table 1 we present figures about the numbers of applied and enrolled students for the study program in the recent three years: 2017, 2018, and 2019, Column (1). According to the enrolment rules at FTS, each applicant, i.e. candidate can express the three options about the study programs wished to study, ordered by the personal candidate's preferences. A principal selection rule in its simplified form is that if a candidate satisfies criteria for his or her Option I, he or she will be offered this option, and removed from the rank lists of the other two options. Otherwise, if a candidate satisfies criteria for Option II, he or she will be offered this option, and removed from the rank list of Option III, etc. Columns (2) and (3) show the number of students applied at B.Sc. in Information Engineering, as Option I, and II and III, respectively. Column (4) shows the total number of applicants, i.e. it is calculated as Column (2) + Column (3). As the quota is 60 students and we have applicants over quota, Column (5) is calculated as Column (4) − 60. According to the enrolment rules at FTS, each candidate has to approach entrance exam. The number of earned points at the entrance exam is from 0 to 60. Also, all candidates bring from secondary school additional points, on the scale from 16 to 40, with respect to their yearly average grades. By this, the maximal number of points earned in the enrolment procedure is $60 + 40 = 100$. A rank list of all applicants for a study program is created in the descending order of the total number of points earned in the enrolment procedure. Column (6) shows the number of points earned by the best ranked candidate(s), while Column (7) shows the number of points earned by the lowest ranked candidate(s) who enrolled the study program.

Table 1. No. of students applied and enrolled at B.Sc. Information Engineering

Year	Option I no.	Options II and III no.	Total no.	Over quota no.	Entrance max. points	Entrance min. points
(1)	(2)	(3)	(4)	(5)	(6)	(7)
2017	46	131	177	117	100.00	69.02
2018	65	215	280	220	91.00	70.42
2019	67	210	277	217	98.24	72.54

As a rule, all CSI&SE study programs at FTS attract much more applicants than there are available places. Typically, a ratio is 4–4.5 candidates per 1 available place.

From Table 1, it follows that the similar ratio is reached from 2018 for the program B.Sc. in Information Engineering, as Column (4)/60 is greater than 4.5. Increasing number of applicants for Option I, as well as for Options II and III indicates a better recognition of this study program by the applicants in 2018 and 2019, than in 2017, as its first year execution. On the other hand side, we find that this study program is still one of predominant Options II and III wishes, on the contrary to the traditional and well established CSI&SE study programs at FTS, where Option I is predominant to Options II and III. This fact, as well as frequent questions of future applicants that we answer before the enrolment procedure, indicate still not clear recognition of study program potentials and its values. We still face many questions by future applicants, as well as current students, about the notion and importance of Data Science in modern, digital society. It is still not clear early in advance to students and applicants, what are the expected knowledge and skill outcomes, after completing B.Sc. in Information Engineering. Furthermore, applicants do not recognize a clear distinction from other CSI&SE programs, particularly from the B.Sc. program in Information Systems Engineering that combines Computer Science and Information Systems with Business Administration. Sometimes, the applicants do not show a clear awareness about a high degree of Mathematics and other formal disciplines included in the study program, while it is a program with much more Mathematics and related disciplines than any other study program at FTS. This fact may be also a reason for predominant Options II and III to Option I in the recent three years of its execution.

In 2017, we also initiated the first execution of the 1-year M.Sc. study program in Information Engineering, and now its third generation of students is in the process. We noticed a strong interest of differently profiled students form B.Sc. level, with programs completed in: Computing and Control, Applied Mathematics /Mathematics, Business Informatics, and Electronics, Telecommunications, and Power Systems. The number of students enrolled per year is still small, from 3 to 10. The similar trend we expect also in the next school year, while the number of interested students will significantly raise in year 2021, after completing the first generation of B.Sc. in Information Engineering.

On the contrary to undergraduate students, graduate students and applicants fully recognize potentials and values of such profiled study program, as well as Data Science in general, and they express their high motivation for studying this discipline. Based on the students' questions and interviews with applicants, we believe that one of the main reasons for a discrepancy between a strong interest for the M.Sc. study program in Information Engineering, and still relatively low number of students being enrolled is in the fact that many M.Sc. students are already employed and feel difficulties in concurrent full time work and studying a demanding subject as Data Science is. Besides, we again face in Serbia or wider region with the problem of a high brain drain. Also, a new fact is that some of students completing their B.Sc. studies decide not to continue with their graduate education, with a belief that B.Sc. level is quite satisfactory for their professional life. All of that reflects the total number of graduate students in engineering study programs.

A positive aspect of M.Sc. study program in Information Engineering is that we recognize a wide selection of Serbian companies, predominantly in software industry, searching for such profiled students. Even, some of them offered quite attractive scholarship schemes in order to support initiating of such study program.

5.3 Study Program Execution

From 2017, we invest our efforts in monitoring students' performances and achievements at the new study program in Information Engineering. We have been collecting students' experiences and opinions mostly from regular meetings that we organize with each generation of undergraduate students. The meetings are organized once in a semester, typically together with the lecturers of all courses in the semester.

In the 2017 generation, we notice a relatively strong polarization 1/2:1/2 of very good over relatively bad students. Typically, students with good performances show noticeably better motivation than the students with bad performances. Even, we notice a strong absence of middle-level students. In this generation, 55 of 60 students in total enrolled Year II, while 50 enrolled Year III. Some of the students enrolled the higher study years even with serious troubles, with not completed all the courses from the previous years. As a rule, such students will come in Year IV to the significant delays in completing their undergraduate studies. In the 2018 and 2019 generations, we notice a solid existence of middle-level students. A distribution of students with good performances over the students with bad performances is slightly better, even to say that it is approximately 2/3:1/3 for good students. However, in 2018 generation, we cannot identify students with excellent performances.

For all undergraduate generations, in the two fundamental courses, Programing Languages and Data Structures, and Theory of Algorithms, we notice a quite strong polarization of students to two clusters. One cluster is of experienced students, and the other is of beginners, in terms of previous programming skills. For the beginners, those courses are declared as very difficult, while for experienced ones, they are declared as easy to normal. Experienced students are almost ready, or even completely ready for independent and individual work in programming, while the beginners need a significant support of teaching assistants, as they are not ready for independent work, at all.

Low half of students' population in Year I feel strong troubles with Algebra in Winter semester. They declare it as a too abstract course, while Calculus 1 is seen as much easier, in Summer semester. We believe that it is due to the fact that many students studied elements of Calculus 1 in their secondary schools, while Algebra is mostly a new subject for majority of them. As it concerns mathematics subjects in Year II, students perceive all of them well: Calculus 2, Graph Theory and Combinatorics, Mathematical Logic, and Probability and Stochastic Processes.

A regular question of many I Year students is about the purpose of the course in Mechanics. They do not believe that it is important subject for them in any way, and they cannot perceive how Mechanics will contribute to their future profession. On the other hand side, they express their quite happiness about the fact that Electrical Engineering courses are not included in the study program. All students perceive well the courses in Computer Architecture, Operating Systems, and Fundamentals of Digital Systems Design. The same is with two fundamental courses in Financial Engineering. They are also particularly happy with the course in Communicology, as they see it as valuable for their general skills and future profession.

Introduction to Information and Financial Engineering is designed as an introductory course to the appropriate disciplines, as well as to Data Science and Information Systems. After completing this course, the students slightly raise their perception of the disciplines of Information Engineering and Data Science. Generally, they perceive the course as not difficult to complete. They are quite familiar with practical exercises and a project assignment that requires developing of various skills: very basic knowledge of SQL and spreadsheet design, data acquisition through a designed questionnaire, basic data analysis and presentation, development of an idea for a new product and business plan that will produce some tangible value, group work, and final group presentation. However, during the lessons, they feel difficulties in understanding the basic notions about the role of Information Systems, Business Intelligence, Data Warehouse, Big Data, and related ones.

As the number of students enrolled in M.Sc. in Information Engineering is still small, our current execution of the program principally includes the following courses: Data Mining and Data Analysis, Data Warehouse Systems and Business Intelligence, Business Process Modeling, Introduction to Interactive Theorem Provers, Quantitative Methods of Risk Management, Cryptography, Statistics in Information Engineering, Domain Specific Modeling and Languages, Professional Practice /Internship, and Master Thesis. Students still have a possibility to replace some of the aforementioned courses with some other ones, and typically they selected some of HPC courses that run in Computing and Control program.

5.4 Reaccreditation of the Study Programs

In 2018, the new 7-year accreditation cycle of all university study programs was initiated in Serbia. On the basis of all collected experiences, we have slightly renovated our B.Sc. and M.Sc. study programs in Information Engineering and Information and Analytics Engineering, and submitted for the accreditation in June 2019. Our two main goals were to embed the best practices of already running students' generations in the study programs, and face with some new or improved regulations set by the National Accreditation Body of Republic of Serbia, in January 2019.

Some of the new requirements set by new regulations were to enlarge the number of hours for mandatory internships, both at undergraduate and graduate levels, as well as the number of hours for B.Sc. and M.Sc. theses.

The most important new requirement for all interdisciplinary programs was a newly introduced metrics, by means of at least 50% of courses and lecturers should come from the first main discipline, while at least 25% of courses and lecturers should come from the second main discipline. In the case of our study programs, the first main discipline is Electrical Engineering and Computing, while the second one is Engineering Management. Criteria and calculation of the aforementioned percentages of course and lecturer participation are not as simple, as it seems to be on the first sight, and we avoid their explanation here. We just communicate here our experience that it was really hard to meet such requirement formally. On the other hand side, a positive aspect of such requirement was that we were forced to even raise the number of elective courses from the two disciplines, and by this, we crated a wider room for our future students to customize their studies to fit better their professional aspirations. In all the three study programs, we raised the number of elective courses, and by this covered even some new disciplines.

One important improvement of the B.Sc. program in Information Engineering is about programming courses. Instead of the courses Programing Languages and Data Structures, and Object Oriented Programming, we introduced the courses Fundamentals of Programming and Programming Languages, and Advanced Programming and Programming Languages, respectively. The main intention is to better design the whole education chain in algorithms and programming, that includes, apart from these two courses, also the course in Theory of Algorithms. By the chain of those three courses, we insist on adopting not only programming skills, but also theoretical knowledge in Programming Languages and Algorithms from the early beginning of university education. Besides, we moved some basics of the object-oriented paradigm to the introductory course in Fundamentals of Programming and Programming Languages, and by this, we released a room in the course Advanced Programming and Programming Languages for some advanced programming paradigms, such as functional, logical, and aspect-oriented paradigm, apart from object-oriented paradigm that is predominantly covered by this course.

Finally, in this reaccreditation, we have slightly moved a ratio of completely newly created courses to 35%, while the rest of 65% are the ones included from other study programs.

6 Conclusion

Following emerging trends about development of Data Science discipline all over the world, development of study programs at recognized universities, as well as our experiences presented in this paper, we conclude that the existence of specific study programs in Data Science is a strong necessity. Faculty of Technical Sciences addressed it by the new study programs, created both at the B.Sc. and M.Sc. levels. Our experiences collected from the program execution in the recent years, show an increasing interest and motivation of students for such discipline. Even the students of the first year of undergraduate studies increase their awareness about the importance of the Data Science discipline in their future career. Also, we find a quite noticeable motivation of education staff to participate in such emerging process of creation and implementation of a high quality study program, and by this develop potentials and giving respectable opportunities to students.

In the future, we are to face with a problem of a polarization of students' population to the one with clear ideas about their future, vs. students with not clear recognition of their future opportunities. We need to address the issue how to further increase awareness of students about the importance of Data Science in upcoming years, and recognize the outcomes and values of the presented study programs.

One of quite characteristic textual comments of our respondents in a completed survey presented in [8] is that "Analytics culture in Serbia is low and organizations do not recognize the importance of BI and how to utilize such data. It is to work on raising organization education, as today business analysis comes down just to operational reporting." On one hand side, our respondents clearly identified strong opportunities for further development in the discipline of Data Science and the expected values of such development, while on the other hand side they reported significant current weaknesses and risks. All of that even more stresses the importance of proactive behavior of university in raising the high level of analytical culture and awareness of Data Science and its role in a data-driven economy, and modern, digital society.

References

1. Anderson, P., Bowring, J., McCauley, R., Pothering, G., Starr C.: An undergraduate degree in data science: curriculum and a decade of implementation experience. In: Proceedings of the 45th ACM Technical Symposium on Computer Science Education, pp. 145–150. ACM, USA (2014). https://doi.org/10.1145/2538862.2538936
2. Chiang, R.H.L., Goes, P., Stohr, E.A.: Business intelligence and analytics education, and program development: a unique opportunity for the information systems discipline. ACM Trans. Manag. Inf. Syst. **3**(3) (2012). Article 12, https://doi.org/10.1145/2361256.2361257
3. Daniel, B.K.: Big data and data science: a critical review of issues for educational research. Br. J. Educ. Tech. **50**(1), 101–113 (2019). https://doi.org/10.1111/bjet.12595
4. Golshani, F., Panchanathan, S., Friesen, O.: A logical foundation for an information engineering curriculum. In: Proceedings of the 30th ASEE/IEEE Frontiers in Education Conference, USA (2000). https://doi.org/10.1109/fie.2000.897639
5. Group of Authors: Informatics education: Europe cannot afford to miss the boat. Report of the Joint Informatics Europe & ACM Europe Working Group on Informatics Education (2013)
6. Klašnja-Milićević, A., Ivanović, M., Budimac, Z.: Data science in education: big data and learning analytics. Comput. Appl. Eng. Educ. **25**(6), 1066–1078 (2017). https://doi.org/10.1002/cae.21844
7. Luković, I.: Formal education in data science – recent experiences from faculty of technical sciences of University of Novi Sad. In: Proceedings of the 21st International Conference Data Analytics and Management in Data Intensive Domains (DAMDID/RCDL 2019), vol. 2523, pp. 19–20, Kazan Federal University, Kazan, Russia. Kazan Federal University, CEUR-WS (2019). ISSN 1613-0073
8. Luković, I.: Formal education in data science – a perspective of Serbia. In: Proceedings of Milićević I. (eds.) 7th International Scientific Conference Technics and Informatics in Education, pp. 12–18, Čačak, Serbia. University of Kragujevac, Faculty of Technical Sciences Čačak, Čačak (2018). ISBN 978-86-7776-226-1
9. Luković, I., Šolaja, M.: Trends of academic education in data science – an analysis of a case of Serbia. In: Proceedings of XXIII Conference Development Trends: A Position of High Education and Science in Serbia (TREND 2017), pp. 162–165, Zlatibor, Serbia. University of Novi Sad, Faculty of Technical Sciences (2017). ISBN 978-86-7892-904-5 (in Serbian)
10. McKinsey Global Institute: Big Data: The Next Frontier for Innovation, Competition, and Productivity. McKinsey, June (2011)
11. Moreira, F., Gonçalves, R., Martins, J., Branco, F., Au-Yong-Oliveira, M.: Learning analytics as a core component for higher education disruption: governance stakeholder. In: Proceedings of the 5th International Conference on Technological Ecosystems for Enhancing Multiculturality, Cádiz, Spain. ACM, USA (2017). https://doi.org/10.1145/3144826.3145387, ISBN 978-1-4503-5386-1
12. Provost, F., Fawcett, T.: Data Science for Business: What You Need to Know about Data Mining and Data-Analytic Thinking. O'Reilly Media (2013). ISBN 978-1-449-36132-7

13. Smith, J.: Data Analytics: What Every Business Must Know about Big Data and Data Science. Pinnacle Publishers, LCC (2016). ISBN 978-1-535-11415-8
14. Tulasi, B.: Significance of big data and analytics in higher education. Int. J. Comput. Appl. **68**(14), 21–23 (2013)
15. Wixom, B., Ariyachandra, T., Goul, M., Gray, P., Kulkarni, U., Phillips-Wren, G.: The current state of business intelligence in academia. Commun. Assoc. Inf. Syst. **29**(1), 299–312 (2011)
16. Yan, D., Davis, G.E.: A first course in data science. J. Stat. Educ. **27**(2), 99–109 (2019). https://doi.org/10.1080/10691898.2019.1623136

Author Index

Printed in the United States
By Bookmasters